WRITTEN IN THE

STARS

CONSTELLATIONS, FACTS AND FOLKLORE

Publishing Director Sarah Lavelle
Editor Harriet Butt
Copy editor Jennifer Latham
Designer Maeve Bargman
Illustrator Jesús Sotés Vicente
Production Director Vincent Smith
Production Controller Tom Moore

Published in 2018 by Quadrille, an imprint of Hardie Grant Publishing

Quadrille
52–54 Southwark Street
London SE1 1UN
quadrille.com

Cataloguing in publication data: a catalogue record for this book is available from The British Library.

Reprinted in 2018, 2019
10 9 8 7 6 5 4 3

ISBN 978 1 78713 176 7

Printed in China

WRITTEN IN THE
STARS

CONSTELLATIONS, FACTS
AND FOLKLORE

ALISON DAVIES

✳

Illustrations by
JESÚS SOTÉS VICENTE

Hardie Grant

QUADRILLE

CONTENTS

INTRODUCTION

Magic sits in the stars. It seeps from the spaces in-between these mysterious orbs, in patches of darkness that serve only to accentuate the glorious sparkle of far-distant planets. In the moment when you catch a glimpse of a special group or pattern of stars in the veil of the sky, for a second you see what the ancients saw. Then you appreciate the magnificence of the universe, and you're taken back, far back, into the annals of time.

Let's not forget, the stars have seen all of life on Earth unfold beneath them. Like eyes, blinking and watching, these tiny, enduring beads of light witness and hold our stories and memories long after we're gone. Since ancient times, astronomers have charted these sparkling pinpricks in the night sky into constellations, giving each one life and meaning, and often a tale to tell. Sky observers have set the constellations neatly in their space and bonded them within their family grouping to give a sense of order to the cosmos. Each pattern of stars slots together like pieces of a giant galactic jigsaw puzzle. And there's no great secret to their success or beauty – they simply are, appearing every night for those who wish to look and experience their wonder.

So let this book be your guide, the springboard from which you dive into this glittering realm. With 88 constellations to traverse, you'll

find much to explore and some tips on when and where to find them in the heavens, what you might see, along with any underlying myths or interesting facts relating to their origins.

Learn to read and appreciate the night sky like the ancients before you, and you'll discover an enchanting world full of surprising insights. Most important of all, you'll understand that, since the beginning of time, the fate of mankind has, and always will be, written in the stars.

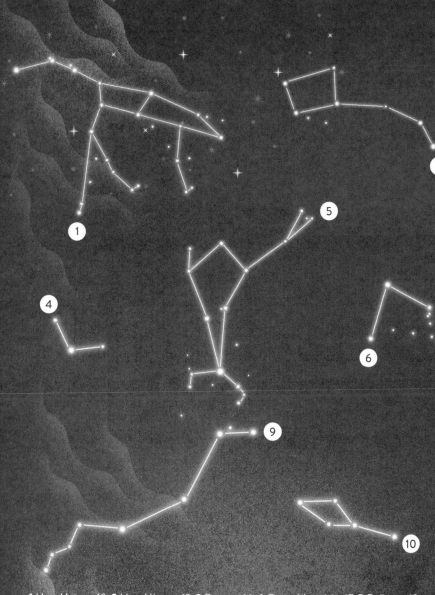

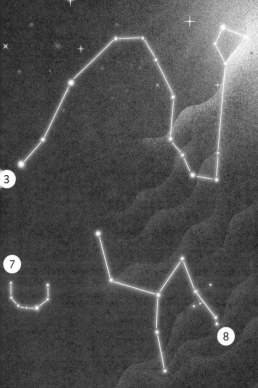

3

7

8

Imagine the scene: a great menagerie of beasts parade before your eyes. Some of them are predatory, others are the prey, and some are animals that once had another form, their fates changed by the gods and then honoured in the stars. Yet others are symbols of beauty and love, cast into the heavens for eternal admiration. The vast family of constellations that make up Ursa Major offers a rich seam of mythology, sprinkled with moral gems that beg closer attention. And like all the best adventures, there's a sprinkling of magic that only the most skilled of storytellers could weave.

URSA MAJOR

URSA MAJOR

◉	**FAMILY**	URSA MAJOR
◐	**LOCATION**	Second quadrant of the northern hemisphere
◉	**BEST SEEN IN**	April
❓	**NAME MEANING**	The Great Bear
⭐	**STARS**	**Alioth** (Epsilon Ursae Majoris), **Dubhe** (Apha Ursae Majoris), **Alkaid** (Eta Ursae Majoris), **Mizar and Alcor** (Zeta Ursae Majoris and 80 Ursae Majoris), **Merak** (Beta Ursae Majoris), **Phecda** (Gamma Ursae Majoris), **Megrez** (Delta Ursae Majoris)

The vast and ever-present Ursa Major is home to the asterism
(a distinct group of stars that are not a constellation) the Plough.
The third-largest constellation in the night sky, Ursa Major's existence
abounds with legends. According to Greek mythology, it represents
a tragic tale of love and envy. Zeus, the king of the gods, seduced the
beautiful nymph Callisto, who bore him a son, Arcas. His wife and
queen, Hera, was enraged and turned the nymph into a bear. Many
years later, Arcas encountered the bear, Callisto, in the forest and
raised his spear in fear. Zeus intervened with an almighty whirlwind,
which sent the she-bear Callisto and Arcas swirling through
the heavens. There Callisto remains, forever bright as
Ursa Major, but an embittered Hera wasn't done.
Her final command was that the bear may not dip
its paws in northern waters, hence Ursa Major
never sets below the horizon in the north.

URSA MINOR

⊙	**FAMILY**	URSA MAJOR
⚡	**LOCATION**	Third quadrant of the northern hemisphere
👁	**BEST SEEN IN**	June
❓	**NAME MEANING**	The Little Bear
★	**STARS**	Polaris (Alpha Ursae Minoris), **Kochab** (Beta Ursae Minoris), **Pherkad** (Gamma Ursae Minoris)

The smaller sibling of Ursa Major, this little bear contains the seven-starred asterism the Little Dipper. According to one Greek legend, this constellation commemorates the nymph Ida, who cared for the infant Zeus on the island of Crete. The story goes that his father, Cronus, the king of the Titans, was so consumed by a prophecy that one of his children would kill him that he set about swallowing his newborn offspring whole (where they stayed until Zeus made Cronus disgorge them). Zeus's mother, Rhea, the Titan queen, saved him from this fate by substituting a stone for the baby Zeus and spiriting him away for his protection. Ursa Minor is home to Polaris, the North or Pole Star, which is situated close to the north celestial pole. It has been used by travellers on land and sea for millennia to find due north, so while it is not the brightest star in the heavens, it is every adventurer's friend.

DRACO

⊙	**FAMILY**	URSA MAJOR
⚡	**LOCATION**	Third quadrant of the northern hemisphere
👁	**BEST SEEN IN**	July
❓	**NAME MEANING**	The Dragon
★	**STARS**	Eltanin (Gamma Draconis), **Athebyne** (Eta Draconis), **Rastaban** (Beta Draconis)

Majestic and awe-inspiring, this sprawling constellation is named after Ladon, the dragon of Greek mythology. Ladon was said to be the offspring of the terrible monster Typhon, and he guarded a golden apple tree in the gardens of the Hesperides, who were nymphs known as the daughters of evening. Legend has it that the fruit conferred immortality on anyone who ate it. Hera gave Ladon the honour of sentry, but didn't count on the great hero Hercules (his popular Roman name, he was called Heracles by the Greeks), who had been charged with stealing one of these precious apples as part of twelve labours set by King Eurystheus. According to one version of the story, Hercules killed Ladon with his poisoned arrows. Hera, being mostly kind of heart, couldn't bear to see the poor dragon in this way, so she cast his image into the stars, where it remains, a glistening sentinel, coiled around the North Pole.

CANES VENATICI

◉	**FAMILY**	URSA MAJOR
⚡	**LOCATION**	Third quadrant of the northern hemisphere
👁	**BEST SEEN IN**	May
❓	**NAME MEANING**	The Hunting Dogs
★	**STARS**	Cor Caroli (Alpha Canum Venaticorum), Chara (Beta Canum Venaticorum), La Superba (Y Canum Venaticorum), AM Canum Venaticorum

Charted in 1687 by the Polish astronomer Johannes Hevelius, this pale dusting of stars falls beneath the tail of Ursa Major. It represents a pair of hunting dogs, Asterion (derived from the ancient Greek word *astéri* meaning 'star') and Chara (the Greek word for 'delight' or 'joy'). They are held tightly on a leash by their master, the herdsman Boötes, in a neighbouring constellation. It's easy to imagine the two hounds in eternal chase, snapping and snarling at the heels of the Great Bear, Ursa Major. While Hevelius was the first to chart the hounds officially in this position, the idea had been around for some time. The 16th-century German astronomer Peter Apian pictured Boötes with two dogs at his heels, although in this case they were following the Herdsman and not worrying the Great Bear.

BOÖTES

⊚	**FAMILY**	URSA MAJOR
⚡	**LOCATION**	Third quadrant of the northern hemisphere
👁	**BEST SEEN IN**	June
❓	**NAME MEANING**	The Herdsman
★	**STARS**	Arcturus (Alpha Boötis), Izar (Epsilon Boötis), Muphrid (Eta Boötis), Nekkar (Beta Boötis), Seginus (Gamma Boötis)

This strapping young herdsman chases Ursa Major through the skies in a dance around the North Pole. With a club in his hand and his hunting dogs, Asterion and Chara, on a leash, he is a fearsome sight. Some ancient Greeks believed he embodied the archetypal ploughman Philomelus driving his oxen forward, but others preferred to think of him as Arcas, the son of Zeus and his mortal lover Callisto. She was turned into a she-bear by a vengeful Hera and became the constellation Ursa Major, the Great Bear, with Arcas tracing his mother's footsteps (his constellation was known to the Greeks at one time as the Bear Keeper). A man with many faces, Boötes in the guise of Philomelus is also credited with having created the first plough (perhaps not surprisingly, as he was said to be the son of Demeter, the goddess of fertility and agriculture), and this earned him his place in the skies.

COMA BERENICES

⊙	**FAMILY**	URSA MAJOR
⚡	**LOCATION**	Third quadrant of the northern hemisphere
👁	**BEST SEEN IN**	May
?	**NAME MEANING**	Berenice's Hair
★	**STARS**	Beta Comae Berenices, Diadem (Alpha Comae Berenices), Gamma Comae Berenices

Named after the Egyptian Queen Berenice II, this scattering of stars hangs softly like the flowing locks of the great beauty. According to legend, the newlywed Queen Berenice was so afraid when her husband, King Ptolemy III Euergetes, went to war, she made a pledge to Aphrodite, the Greek goddess of love. If the goddess used her powers to bring him back safely, Berenice would cut off her golden tresses in a show of gratitude. When her husband returned in one piece, Berenice honoured her word. She cut off her hair, and left it in Aphrodite's temple. The next day, her locks had disappeared, much to the annoyance of her husband. To soothe his temper, the court astronomer told him that Aphrodite had been so enamoured with the hair, she'd spread it in the sky for all to see its natural splendour.

CORONA BOREALIS

◎	**FAMILY**	URSA MAJOR
ⓕ	**LOCATION**	Third quadrant of the northern hemisphere
◉	**BEST SEEN IN**	July
❓	**NAME MEANING**	The Northern Crown
★	**STARS**	Alphecca (Alpha Coronae Borealis), Nusakan (Beta Coronae Borealis), Gamma Coronae Borealis

Rich in mythology, this dazzling array of stars is thought to be the jewelled wedding crown worn by Ariadne, princess of Crete, famous for her role in the Greek legend of Theseus and the Minotaur. After her marriage to Dionysus, the god of wine, the deity tossed her crown into the sky, where it remains, a glistening symbol of love. Various Native American tribes named it the Camp Circle, a starry reflection of their camps below. The Celts believed this grouping of stars was a mystical place known as Caer Arianrhod, or the Castle of the Silver Circle, a place of great power and home of the mythical Lady Arianrhod, sometimes referred to as the goddess of the moon and the stars.

CAMELOPARDALIS

⊙	**FAMILY**	URSA MAJOR
⚡	**LOCATION**	Second quadrant of the northern hemisphere
👁	**BEST SEEN IN**	February
❓	**NAME MEANING**	The Giraffe
★	**STARS**	Beta Camelopardalis, CS Camelopardalis, Alpha Camelopardalis, Struve 1694 (Sigma 1694 Camelopardalis)

This constellation may be faint, but its name has a charming origin. The ancient Greeks thought the giraffe was a fantastical creature, the result of a pairing between a camel (*kamelos* in ancient Greek) and a leopard (*pardalis*), hence 'camel-leopard', an animal with a slender neck, long legs and a body littered with spots. The first giraffe to be seen in Europe was brought much later by the Roman general Julius Ceasar in 46 BC. Scholars agree there's little mythological reference for this constellation, and it was created by the Dutch astronomer Petrus Plancius in the 17th century. However, the German astronomer Jakob Bartsch described it in his star atlas of 1624, probably in a misunderstanding, as the camel Rebecca rode into Canaan on her journey to wed Isaac in the Book of Genesis. Rebecca was chosen to be Isaac's bride after an act of kindness in which she brought water from a well to his father's thirsty camels.

LYNX

⊙	**FAMILY**	URSA MAJOR
⚡	**LOCATION**	Second quadrant of the northern hemisphere
⊙	**BEST SEEN IN**	March
❓	**NAME MEANING**	The Lynx
★	**STARS**	Alpha Lyncis, 38 Lyncis, 10 Ursae Majoris, Alsciaukat (31 Lyncis)

Although it's not the brightest constellation in the night sky, the Lynx gets its name from one of this creature's keenest attributes. Astronomer Johannes Hevelius, who first introduced this constellation in 1687, claimed the eyesight of a lynx was essential for those who wanted to gaze upon its wonder. Some scholars have linked the Lynx to Lynceus, a character from the ancient Greek tales of the hero and adventurer Jason and his heroic shipmates, the Argonauts, in their quest for the fabled Golden Fleece. Thought to have the most impressive eyesight in all the world, Lynceus could see long distances, in the dark and beneath the ground, making him an essential companion during the Argonauts' long expedition.

LEO MINOR

⊙	**FAMILY**	URSA MAJOR
⚡	**LOCATION**	Second quadrant of the northern hemisphere
⦿	**BEST SEEN IN**	April
❓	**NAME MEANING**	The Lesser Lion
★	**STARS**	Praecipua (46 Leonis Minoris), Beta Leonis Minoris, 21 Leonis Minoris, RY Leonis Minoris

Though it may be tiny, this lion cub still has a glorious sparkly mane and something of a playful nature. It was created in 1687 by astronomer Johannes Hevelius from the 18 stars that twinkle between Ursa Major and Leo. He depicts it in his star atlas as though about to catch the tip of Lynx's tail, a neighbouring constellation. Leo Minor constellation was renamed Leaena, or the Lioness, in 1870 by English astronomer Richard A. Proctor in a bid to shorten its name. The new moniker didn't catch on, and it remains to this day the small but proud Leo Minor.

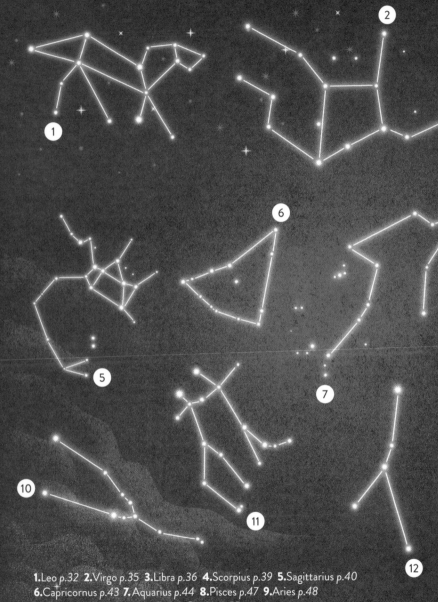

You don't need to be an avid fan of horoscopes to enjoy this family of constellations. While the names may be familiar, the narratives offer a surprising insight into the hearts and minds of the ancients. Servants of the gods in most cases, each zodiac arrangement tells a deeper story. There's love, honour, courage and bitter betrayal, and every mythical character has a part to play. With the canopy of the night sky as a movie screen, and the shape of the stars bringing each myth to life, it's easy to see how astrologers have forged a much stronger connection to the signs of the zodiac – simply by looking up.

ZODIAC

LEO

⊙	**FAMILY**	ZODIAC
⚡	**LOCATION**	Second quadrant of the northern hemisphere
👁	**BEST SEEN IN**	April
❓	**NAME MEANING**	The Lion
★	**STARS**	**Regulus** (Alpha Leonis), **Algieba** (Gamma Leonis), **Denebola** (Beta Leonis)

The majestic lion stalks the heavens, and as befits such a noble creature he is accompanied by some of the brightest stars in the night sky. An enormous beast known as the Nemean lion, according to Greek legend the animal could not be killed because of its magical pelt, which prevented any weapon from piercing it. The lion was the scourge of the countryside, preying on the people and leaving a trail of bloodshed. The first of his famous twelve labours set by King Eurystheus, Hercules was instructed to destroy the creature. He tried at first to kill the lion with his bow and arrows, but when this failed he used his mighty strength to choke it to death. Hercules is often depicted wearing its skin, complete with its claws, as a cloak to mark his victory. One of the oldest constellations in the night sky, it was also known as the Lion by the Syrians and the Turks.

VIRGO

◉	**FAMILY**	ZODIAC
⚡	**LOCATION**	Third quadrant of the southern hemisphere
👁	**BEST SEEN IN**	May
❓	**NAME MEANING**	The Maiden
★	**STARS**	Spica (Alpha Virginis), **Porrima** (Gamma Virginis), **Vindemiatrix** (Epsilon Virginis)

With angel-like wings, this beautiful woman captivates from her place in the heavens. She holds an ear of corn in one hand, marked by the luminous star Spica (the Latin word for an ear of wheat). There are two key Greek mythological figures associated with this constellation. The first is Dike, the virgin goddess of justice and righteousness, and one of three sisters (the Horae) known as the goddesses of the seasons. She once lived on earth, an immortal queen among mortals, during the Golden Age of mankind. This was a time of peace, harmony and bountifulness when spring ruled eternal. Then the Bronze Age dawned and men forged swords and waged war. When Dike's pleas to turn their backs on such violent and lawless ways fell upon deaf ears, she flew to the heavens to be with the other deities. The second figure is Demeter, the goddess of agriculture, who is often depicted holding sheaves of wheat.

LIBRA

◉	**FAMILY**	ZODIAC
⚡	**LOCATION**	Third quadrant of the southern hemisphere
◉	**BEST SEEN IN**	June
❓	**NAME MEANING**	The Scales
⭐	**STARS**	Zubeneschamali (Beta Librae), Zubenelgenubi (Alpha Librae)

Known as the Balance of Heaven by the ancient Sumerians, Libra represents the pair of scales held by Dike, the Greek goddess of justice, which are sited at the feet of her constellation. Its four brightest stars form a quadrangle, and it also harbours the oldest star in the night sky known as Methuselah. Named after the Old Testament patriarch who was said to have lived for 969 years, scientists currently estimate that this star is around 14.5 billion years old – that's older than the Big Bang 13.8 billion years ago! The Romans believed that the Moon was firmly in Libra when the great city of Rome was first founded (legend has it that this was on 21 April 753 BC). This constellation also represented the balance of the seasons for the Romans, as the sun passed through the autumn equinox in Libra at that time, a season when the length of day and night are equal.

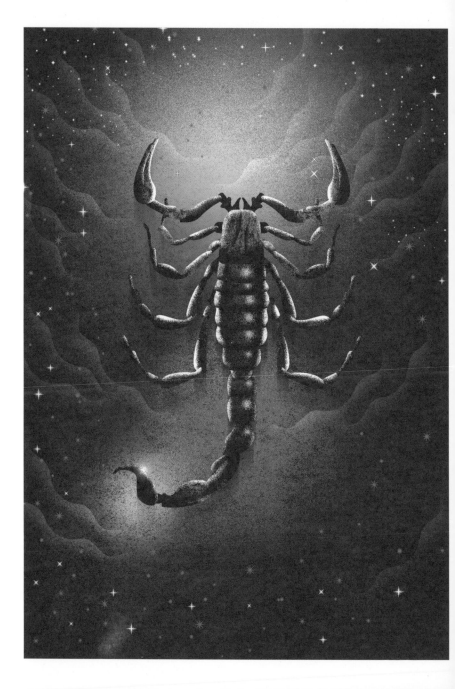

SCORPIUS

⊙	**FAMILY**	ZODIAC
⚡	**LOCATION**	Third quadrant of the southern hemisphere
👁	**BEST SEEN IN**	July
❓	**NAME MEANING**	The Scorpion
★	**STARS**	Antares (Alpha Scorpii), **Shaula** (Lambda Scorpii), **Sargas** (Theta Scorpii)

A giant scorpion crawls through the heavens in search of the hunter Orion. According to one version of the Greek myth, Scorpius was sent by the beautiful Artemis (she is better known by her Roman name of Diana), the goddess of the hunt and wild animals, to strike the hero down because he tried to ravish her; eventually the arachnid dealt the hunter a fatal sting. (A different account claims that Artemis killed Orion herself with an arrow.) Yet another narrative suggests Orion bragged he could kill any animal on earth, which angered Gaia, the goddess of the earth, who sent the scorpion in hot pursuit to punish him. After the deadly deed was done, it's said that Zeus placed Scorpius in the sky to remind men that they should not be too boastful.

SAGITTARIUS

◎	**FAMILY**	ZODIAC
⚡	**LOCATION**	Fourth quadrant of the southern hemisphere
👁	**BEST SEEN IN**	August
❓	**NAME MEANING**	The Archer
★	**STARS**	Kaus Australis (Epsilon Sagittarii), Nunki (Sigma Sagittarii), Ascella (Zeta Sagittarii), Kaus Media (Delta Sagittarii), Kaus Borealis (Lambda Sagittarii), Pistol Star (V4647 Sagittarii)

The Centaur stands, arrow raised in the direction of the Scorpion, a powerful hybrid of man and horse. Some ancient Greek astronomers argued that this constellation represented the satyr Crotus, who was a nurse and companion to the nine Muses, the goddess daughters of Zeus and patronesses of the arts, who lived on Mount Helicon. A mythical creature with two goat-like legs and tail, Crotus was the son of Pan, who among other things was recognised as the god of shepherds and hunters. Known for inventing archery and for his prowess at hunting and horseback riding, when Crotus died the Muses requested that his image be cast into the sky. The Babylonians had a different view. They believed the centaur represented Pabilsag, a winged god with various attributes, including that of war and hunting. Pabilsag has also been identified with Nergal, the god of plague and pestilence.

CAPRICORNUS

◉	**FAMILY**	ZODIAC
⚡	**LOCATION**	Fourth quadrant of the southern hemisphere
◉	**BEST SEEN IN**	September
❓	**NAME MEANING**	The Sea Goat
★	**STARS**	Deneb Algedi (Delta Capricorni), **Dabih** (Beta Capricorni), **Algiedi** (Alpha Capricorni)

Although this may be one of the faintest constellations in the night sky, Capricornus dates back to around 1000 BC. Both the Babylonians and the Sumerians knew this arrangement of stars as the goat-fish. Marking the start of the winter solstice during the early Bronze Age, Capricornus was a practical constellation. In Greek and Roman mythology, it was associated with the god Pan, or possibly with his alter ego, a Panes (rustic spirit or faun) called Aegipan. The ancients believed that he won his place in the sky by helping the other gods on several occasions. He is said to have caused *panikos* (panic) in the Titans when they were at war with the Olympian gods, and amusingly he also threw shellfish at them, hence one possible explanation for his fishy posterior. He also helped defeat the terrible monster Typhon. A thankful Zeus decided to create his image in the stars as a way of remembering his deeds.

AQUARIUS

◉	**FAMILY**	ZODIAC
⚡	**LOCATION**	Fourth quadrant of the southern hemisphere
👁	**BEST SEEN IN**	October
❓	**NAME MEANING**	The Water Bearer or Cup Bearer
⭐	**STARS**	Sadalsuud (Beta Aquarii), Sadalmelik (Alpha Aquarii), Skat (Delta Aquarii)

Aquarius stands ankle-deep in the sea of the firmament. This area of the sky is known as the 'sea' because most of the constellations close by are associated with the ocean in some way. The ancient Greeks claimed that Aquarius represented a handsome prince of Troy known as Ganymede, who was plucked up by Zeus because of his favourable face to serve the gods as cupbearer on Mount Olympus, the home of the Olympian gods. Aquarius can be seen pouring water into the mouth of Pisces Austrinus, the Southern Fish. To the ancient Egyptians this figure also had watery and life-giving connotations. He represented Hapi, the god of the Nile, who each year poured the water from his urn into the great river for the annual summer floods. Thus the banks would overflow, distributing a rich silt into the Nile Valley and ensuring the fertility of the land.

PISCES

◉	**FAMILY**	ZODIAC
🗲	**LOCATION**	First quadrant of the northern hemisphere
👁	**BEST SEEN IT**	November
❓	**NAME MEANING**	The Fishes
⭐	**STARS**	Kullat Nunu (Eta Piscium), **Gamma Piscium**, Alrescha (Alpha Piscium)

To the Babylonians, this vast constellation represented two fish joined together by a cord, swimming in the ocean. The Romans had a similar interpretation; they believed that the fish were Venus, the goddess of love and beauty, and her charming son Cupid, the god of love. The story is set during the war of the gods against the hideous winged monster Typhon. A giant with glaring red eyes, his shoulders writhed with a mass of fiery-eyed, viperous snakes that let out awful hissings, bellows and roars, and his favoured method of attack was to hurl red-hot rocks. Aphrodite and Cupid were walking in the forest when they heard the monster approaching. Terrified for their lives, they leapt into a nearby river and turned into fish, having the foresight to tie themselves together with rope first so they would not lose each other. The ancient Greeks believed this to be a fortunate constellation as its appearance heralded the finer weather of spring.

ARIES

◉	**FAMILY**	ZODIAC
⚡	**LOCATION**	First quadrant of the northern hemisphere
👁	**BEST SEEN IN**	December
❓	**NAME MEANING**	The Ram
★	**STARS**	Hamal (Alpha Arietis), Sheratan (Beta Arietis), Bharani (41 Arietis), Botein (Delta Arietis)

While this arrangement of stars may not look like a ram, Aries takes centre stage across the ecliptic, making it an important constellation to the ancient Greeks. It's thought to represent a flying ram with a golden fleece who tried to save the twins Phrixus and Helle, children of the cloud nymph, from certain death. Their *stepmother* Ino hated them and, through various machinations devised that they should be sacrificed to the gods. Nephele sent the ram to the rescue as Phrixus lay on the sacrificial altar; Phrixus survived the flight, but Helle fell off the beast and drowned before reaching safety. Later the ram was sacrificed to Zeus, and Phrixus gave the precious fleece to King Aeëtes as a thank you for taking him under his protection. The gleaming fleece was hung from a large oak tree in the garden of Ares, a symbol of power and authority, and was guarded by a dragon-serpent that never slept.

TAURUS

◎	**FAMILY**	ZODIAC
⚡	**LOCATION**	First quadrant of the northern hemisphere
👁	**BEST SEEN IN**	January
❓	**NAME MEANING**	The Bull
★	**STARS**	Aldebaran (Alpha Tauri), **Elnath** (Beta Tauri), **Alcyone** (Eta Tauri), **Pectus Tauri** (Lambda Tauri), **119 Tauri**

Known as the Heavenly Bull to the Babylonians, Taurus the Bull has been worshipped as a sacred animal by many cultures over the millennia. The constellation is commonly associated with Zeus, who, according to Greek myth, shapeshifted into a handsome yellow bull in order to charm the beautiful princess Europa. The plan worked, and Zeus the bull carried her off across the ocean to the island of Crete, where he made her his bride. The constellation of Taurus is also associated with the bull-headed, flesh-eating Minotaur, who lived in a winding labyrinth on Crete and was killed by the Greek hero Theseus. In the *Epic of Gilgamesh*, a literary work from ancient Mesopotamia, the Bull of Heaven descends to earth to face the hero of the same name. He was sent by the goddess Ishtar because Gilgamesh spurned her lustful advances – proof of the old adage 'hell hath no fury like a woman scorned' – but fortunately Gilgamesh kills the beast.

GEMINI

⊙	**FAMILY**	ZODIAC
⚡	**LOCATION**	Second quadrant of the northern hemisphere
👁	**BEST SEEN IN**	February
❓	**NAME MEANING**	The Twins
★	**STARS**	Pollux (Beta Geminorum), Castor (Alpha Geminorum), Alhena (Gamma Geminorum)

Known as the Dioscuri, which means 'the sons of Zeus', the twins Castor and Pollux are identified with this constellation. Despite being twins, they did not have the same father. Castor was the son of the king of Sparta, while Pollux was the son of Zeus (not so outlandish as it sounds, as, thanks to DNA testing, it's now known that, although very rare, twins can have separate fathers, when two ova are fertilized by two different men). Their mother, Leda, was seduced by Zeus in the guise of a swan and she fell pregnant with Pollux. Both twins were superb warriors, among other adventures accompanying Jason and the Argonauts. When the Argonauts were caught in a violent storm, the crew prayed for deliverance. Immediately the storm stopped and a light shone above the twins' heads: a sign from the gods. As a result, they were patron gods of sailors, blessed with the power to help those in trouble at sea.

CANCER

⊙	**FAMILY**	ZODIAC
⚡	**LOCATION**	Second quadrant of the northern hemisphere
👁	**BEST SEEN IN**	March
❓	**NAME MEANING**	The Crab
⭐	**STARS**	Al Tarf (Beta Cancri), Asellus Australis (Delta Cancri), Acubens (Alpha Cancri)

This valiant crab, known by the ancient Greeks as Carcinus, was sent by Hera to plague Hercules during his battle with the serpent Hydra. The giant crab tried in vain to fulfil his mission, snapping at Hercules' ankles, but was crushed by the hero's foot in the struggle. Hera took pity on the creature, tossing it high into the sky so that its image would remain a permanent reminder of its efforts. The ancient Egyptians associated the constellation of Cancer with the sacred scarab beetle, also known as the dung beetle. This humble insect spends most of its life rolling dung into balls, which provide its primary food source and a place to lay its eggs. To the Egyptians this symbolised both the action of Khepri, the god of the morning sun, as he rolled out the sun each day, and of rebirth and renewal.

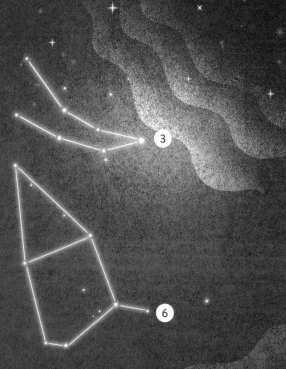

Love has always been a powerful motivating force. To the ancient Greeks, it was the essence of all life, a magical energy that could be harnessed and used in the execution of incredible feats and displays of heroism. The constellations in this family depict different kinds of love, from the self-centred love of a hedonistic queen, to the selfless unconditional love of a daughter who is willing to put her life on the line for her people. There's romantic love too, and that emotion that is so often inspired by beings of beauty: a sense of joy at the spark of creation. On reading the tales, it's easy to see how these stars, lovingly charted by the ancients, won their rightful place in the night sky.

P E R S E U S

CASSIOPEIA

◉	**FAMILY**	PERSEUS
⚡	**LOCATION**	First quadrant of the northern hemisphere
👁	**BEST SEEN IN**	November
❓	**NAME MEANING**	The Seated Queen
⭐	**STARS**	Gamma Cassiopeiae, Schedar (Alpha Cassiopeiae), Caph (Beta Cassiopeiae), Ruchbah (Delta Cassiopeiae), Segin (Epsilon Cassiopeiae)

The stunning yet arrogant Queen Cassiopeia was married to King Cepheus of Ethiopia. As vain as she was magnificent, she boasted that she was more beautiful than the Nereids, a group of sea nymphs, which plunged her into deep and precarious waters. Some variations of the Greek myth claim she was boasting about her daughter Andromeda, or even about both of them! An unimpressed Poseidon, who ruled the oceans, sent the sea monster Cetus to destroy the coast of Cepheus's kingdom. A local oracle advised Cepheus that the only way to appease the monster was to sacrifice Andromeda to it. Reluctantly Cepheus did so, chaining her to the rocks where she was saved by the hero Perseus in the nick of time. Cassiopeia's heartless vanity remained throughout and she was tied to her throne and condemned to the skies where she orbits the North Pole for eternity, and spends half of the year hanging upside down from her starry prison.

CEPHEUS

◉	**FAMILY**	PERSEUS
⚡	**LOCATION**	Fourth quadrant of the northern hemisphere
👁	**BEST SEEN IN**	October
❓	**NAME MEANING**	The King
★	**STARS**	Alderamin (Alpha Cephei), Alfirk (Beta Cephei), Errai (Gamma Cephei), Delta Cephei, Herschel's Garnet Star (Mu Cephei)

The stunning Herschel's Garnet Star, one of the largest and brightest in the Milky Way, resides in this constellation, which is named in Greek myth after King Cepheus of Ethiopia, a land that lay at the Mediterranean tip of the Red Sea. His queen was the vain but beautiful Cassiopeia. Cepheus is most famous for consulting the Oracle of Ammon when his land was under attack from the sea monster Cetus. The oracle told him the only way to save his kingdom would be to offer up his daughter, the lovely Andromeda, to the ravening beast. The story has a happy ending as Andromeda was saved from the monster by the hero Perseus.

ANDROMEDA

◉	**FAMILY**	PERSEUS
⚡	**LOCATION**	First quadrant of the northern hemisphere
👁	**BEST SEEN IN**	November
❓	**NAME MEANING**	The Chained Maiden
★	**STARS**	Alpheratz (Alpha Andromedae), Mirach (Beta Andromedae), Almach (Gamma Andromedae)

Andromeda is celebrated in Greek myth as the beautiful Ethiopian princess whose parents, King Cepheus and Queen Cassiopeia, had her chained to a rock as a sacrifice to the grisly sea monster Cetus in the hope that this would stop him from ravaging their shores. Indeed, it's said that the virtuous Andromeda was willing to give her life for her kingdom. Luckily, the hero Perseus happened to fly by on his winged sandals and, overcome by her beauty and moved by her plight, offered to slay the monster in return for her hand in marriage. Andromeda's parents readily agreed, and a great battle took place between Perseus and Cetus, with Perseus the victor. Andromeda and Perseus later went on to wed and had seven children. After Andromeda's death, the feisty Athena, the goddess of wisdom and warfare and patroness of heroic endeavours, placed Andromeda's image among the stars in a V-shaped constellation to honour the princess for her selfless deed.

PERSEUS

◉	**FAMILY**	PERSEUS
⚡	**LOCATION**	First quadrant of the northern hemisphere
👁	**WBEST SEEN IN**	December
❓	**NAME MEANING**	The Hero
⭐	**STARS**	Mirfak (Alpha Persei), Algol (Beta Persei), Atik (Zeta Persei)

Named after the legendary Greek hero Perseus, this large constellation lays claim to Algol, a bright-red star said to represent the head of the Gorgon Medusa, a monstrous winged female with snaky hair and a face that turned any living being who looked upon her to stone. Perseus famously killed Medusa by using his shield as a mirror to avoid her direct gaze; he lopped off her head and used it to turn many of his foes to stone. The son of Zeus, Perseus features in numerous myths, including that of Andromeda and the sea monster Cetus (also nearby constellations). A great warrior and intrepid adventurer, he is most often depicted wearing a helmet that made him invisible, winged sandals and carrying a jewelled sword in one hand and Medusa's terrifying head in the other. It's no surprise, therefore, that this hero of the heavens boasts a stunning meteor shower, the Perseids, that rivals any other in the night sky.

PEGASUS

◎	**FAMILY**	PERSEUS
⚡	**LOCATION**	Fourth quadrant of the northern hemisphere
👁	**WHEN TO SEE IT**	October
❓	**NAME MEANING**	The Winged Horse
★	**STARS**	Enif (Epsilon Pegasi), Scheat (Beta Pegasi), Markab (Alpha Pegasi)

Like a glorious starburst, Pegasus was born when Perseus beheaded the Gorgon Medusa, according to Greek myth. A beautiful white-winged stallion, he sprang from Medusa's severed neck and fled to Mount Helicon, the sacred home of the Muses, the goddesses of the arts. His name is derived from the Greek *pege* (plural *pegai*) which means 'spring', 'spring-fed well' or 'fountain', and it's said that he created the glistening spring Hippocrene, also known as the horse's fountain, by striking his hoof against a rock. Those who drank from this magical spring were blessed with inspiration – not unlike those who gaze at this wondrous constellation, the seventh largest in the night sky. Pegasus was later tamed by the Greek hero Bellerophon and helped him destroy the fire-breathing monster Chimera. He then served Zeus by carrying his thunderbolts, thus earning his place among the stars.

CETUS

◎	**FAMILY**	PERSEUS
⚡	**LOCATION**	First quadrant of the southern hemisphere
👁	**BEST SEEN IN**	December
❓	**NAME MEANING**	The Sea Monster
⭐	**STARS**	Diphda (Beta Ceti), Menkar (Alpha Ceti), Mira (Omicron Ceti), Tau Ceti

Cetus is the sea monster in the Greek myth of Andromeda and Perseus. Sent to ravage the coastline by Poseidon after Andromeda's mother, Cassiopeia, offended the Nereid sea nymphs by claiming she would always win in a beauty contest with them, he was eventually killed by Perseus. Although it's thought he was a whale, in the myth the description differs. It suggests he was more like a giant sea dragon with terrifying jaws, clawed forefeet and a fish's tail. Whatever the truth, his size is the one common denominator, and just like the constellation he represents – one of the largest in the sky – he commands the heavens.

AURIGA

⊙ **FAMILY** PERSEUS

⚡ **LOCATION** First quadrant of the northern hemisphere

👁 **BEST SEEN IN** February

❓ **NAME MEANING** The Charioteer

★ **STARS** Capella (Alpha Aurigae), **Menkalinan** (Beta Aurigae), **Mahasim** (Theta Aurigae)

Named after the pentagonal shape of the stars that form the pointed peak of a charioteer's helmet, Auriga is linked to Erichthonius, a Greek hero who became king of Athens. He was the product of a mysterious encounter between Hephaestus, the god of blacksmiths and fire (known by the Romans as Vulcan), and the goddess Athena, who raised him as her son. Possibly the original horse whisperer, Erichthonius was famed for his herd of 3,000 horses and for his ability to tame and harness the wild steeds. The creator of the four-horse chariot, his impressive talents won the respect of Zeus, who later placed him in the sky, a shining constellation for all to admire. He is often depicted holding reins in one hand and carrying a goat and two kids in the other. The bright star Capella, known as the she-goat, sits in this constellation, and possibly represents the goat Amalthea. She nursed Zeus as a baby, along with her two kids who are identified with the stars Eta and Zeta Aurigae.

LACERTA

⊙	**FAMILY**	PERSEUS
⚡	**LOCATION**	Fourth quadrant of the northern hemisphere
⊙	**BEST SEEN IN**	October
❓	**NAME MEANING**	The Lizard
★	**STARS**	Alpha Lacertae, 1 Lacertae, 5 Lacertae

A delicate constellation forming the shape of a tiny W, similar in form to the constellation of Cassiopeia, Lacerta's nickname is Little Cassiopeia. It was named by the astronomer Johannes Hevelius in 1687, and originally called Stellio after the starred agama, a type of lizard with rows of fine spines that run along its body, tail and neck. Over the years the name morphed into the much simpler Lacerta or Lizard. Interestingly, the Chumash people of California, who were keen astronomers, also referred to this area of the heavens as the Lizard.

TRIANGULUM

⊙	**FAMILY**	PERSEUS
⚡	**LOCATION**	First quadrant of the northern hemisphere
👁	**BEST SEEN IN**	December
❓	**NAME MEANING**	The Triangle
⭐	**STARS**	Beta Trianguli, Mothallah (Alpha Trianguli), Gamma Trianguli, HD 13189

This diminutive constellation takes the shape of a triangle, hence its name. The ancient Greeks knew it as Deltoton, as its form was likened to the Greek capital letter delta (Δ). The Romans called it Sicilia after the island of Sicily, which is roughly triangular in shape. Ceres, the nurturing mother goddess of Roman mythology, who is associated with the land and agriculture, is also linked to this island and was Sicily's patron goddess. It's thought she implored Jupiter, the Roman king of the gods and god of the sky, to cast its form into the firmament, where it would twinkle forever.

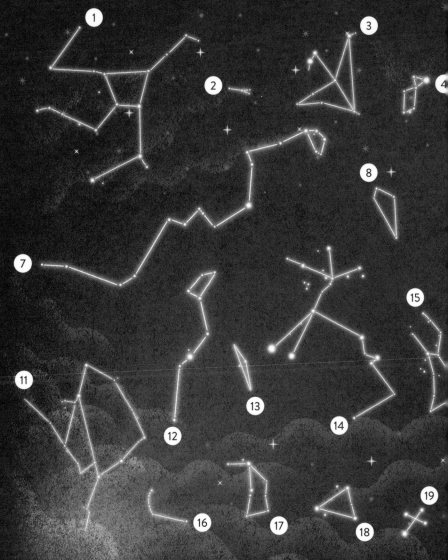

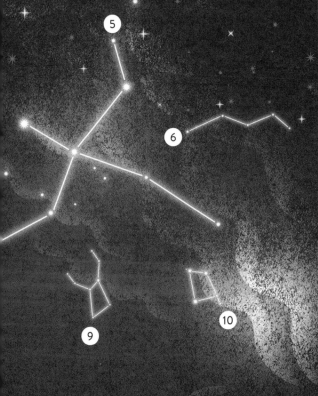

The ancients knew how to celebrate their heroes. Whatever shape or form they took, they were revered and often cast into the stars to be immortalised and worshipped forever. Mighty Hercules was one such hero, appearing in both Greek and Roman myth. Many of his trials and tribulations are charted in the heavens with tales of dragons, fearsome eagles and multi-headed water serpents, all nestled beneath his powerful shoulders. In his family you'll find tales of wit and wonder, of great love and friendship too: stories that are meant to inspire and ignite the heart. Just as these constellations illuminate the sky, they can also shed light on our world today.

HERCULES

HERCULES

⊙	**FAMILY**	HERCULES
⚡	**LOCATION**	Third quadrant of the northern hemisphere
👁	**BEST SEEN IN**	July
❓	**NAME MEANING**	Third quadrant of the northern hemisphere
★	**STARS**	Kornephoros (Beta Herculis), Zeta Herculis, Sarin (Delta Herculis)

These stars outline the shape of a man kneeling over the body of a slain dragon. One thing is clear – he is a hero. This constellation charts the mighty Hercules of Greek myth in his moment of triumph over the dragon Ladon, one of the twelve labours set by King Eurystheus. A tragic tale lies behind this legend. Although Hercules was a son of Zeus and the mortal Alcmene, renowned as the strongman of the ancient world, his life was cursed by Zeus's wife, Hera, who, unable to punish her husband, vowed to make Hercules' life a misery instead. On one occasion she drove Hercules in a moment of madness to kill his wife and children. Horrified by this terrible deed, Hercules sought to atone for his actions and was told by the Oracle of Delphi to serve King Eurystheus. Although a famous Greek myth, it has its roots in Sumerian folklore, and Hercules is often compared to the Sumerian hero Gilgamesh.

SAGITTA

⊙	**FAMILY**	HERCULES
⚡	**LOCATION**	Fourth quadrant of the northern hemisphere
👁	**BEST SEEN IN**	September
❓	**NAME MEANING**	The Arrow
★	**STARS**	Gamma Sagittae, Delta Sagittae, Sham (Alpha Sagittae)

This constellation may be tiny, the third smallest in the sky, but its namesake, the Arrow, achieved mighty deeds. According to the ancient Greeks, this sharp and feisty weapon belonged to the hero Hercules, who used it to kill the eagle that daily gnawed on the liver of the great Titan Prometheus. Each night his liver re-grew, only to be pecked on again the next day, dooming Prometheus to perpetual torment. This cruel fate was arranged by Zeus to punish him for giving mankind the gift of fire. Other stories assert that the arrow belonged to Eros, the god of love, and was used many times to set hearts fluttering. It's also thought it was the same arrow that Apollo, the sun god, deftly used to kill the Cyclopes for their involvement in the death of his son Asclepius. Either way, the arrow, though small, plays an important role in folklore, just as this constellation has earned its place in the sky.

AQUILA

◉	**FAMILY**	HERCULES
⚡	**LOCATION**	Fourth quadrant of the northern hemisphere
👁	**BEST SEEN IN**	September
❓	**NAME MEANING**	Fourth quadrant of the northern hemisphere
★	**STARS**	Altair (Alpha Aquilae), **Tarazed** (Gamma Aquilae), **Deneb el Okab Australis** (Zeta Aquilae)

According to Greek myth, at the start of the war between the giant race of Titans and the Olympian gods, Zeus saw an eagle fly overhead and read this as a good omen. He made the eagle the king of all birds, and honoured the raptor with guarding his great sceptre. This winged king of the heavens was also entrusted with carrying thunderbolts for Zeus. Other myths suggest he's the eagle that carried the beautiful youth Ganymede to Mount Olympus, the home of the gods, to be their cupbearer. While there are many stories attached to this constellation, it's evident the eagle was a symbol of power and that Aquila was an important part of the celestial vista. Both the Babylonians and the Sumerians noted the way its brightest star, Altair, commanded the heavens, calling it the eagle star.

LYRA

⊙	**FAMILY**	HERCULES
ⓘ	**LOCATION**	Fourth quadrant of the northern hemisphere
◉	**WHEN TO SEE IT**	August
❓	**NAME MEANING**	The Lyre
★	**STARS**	Vega (Alpha Lyrae), **Sulafat** (Gamma Lyrae), Sheliak (Beta Lyrae)

Named after the golden lyre presented to the musician Orpheus by Apollo, Lyra is a small constellation. The lyre of Greek myth produced beautiful music, so captivating it could charm the birds from the trees and the stones on the ground. Famously, Orpheus accompanied Jason and the Argonauts and used his lyre to drown out the calls of the Sirens, who lured sailors onto perilous rocks, ensuring the Argonauts' safe passage. When his beloved wife, Eurydice, died, Orpheus used his lyre to charm his way into the Underworld to ask Hades, the god of the dead, to release her. He agreed, but only if Orpheus promised not to gaze on her until safely home. Unfortunately anxious Orpheus couldn't resist a look, and in a trice Eurydice returned to Hades. When Orpheus eventually met his demise, his lyre was cast into a river. Zeus sent an eagle to reclaim it, then placed both the eagle and the lyre in the stars.

CYGNUS

◉	**FAMILY**	HERCULES
⚡	**LOCATION**	Fourth quadrant of the northern hemisphere
👁	**WHEN TO SEE IT**	September
❓	**NAME MEANING**	The Swan
★	**STARS**	**Deneb** (Alpha Cygni), **Sadr** (Gamma Cygni), **Aljanah** (Epsilon Cygni), **Albireo** (Beta Cygni)

The most poignant tale associated with this constellation is that of the mortal Phaethon, son to the charioteer of the sun, and his friend Cycnus. Foolish Phaethon took the sun chariot for a spin but couldn't control it, so Zeus sent a thunderbolt that plunged the youth to his death in the River Eridanus. Cycnus, unable to free his friend's body from its underwater grave, implored Zeus to transform him into a swan. In return he would only have the lifespan of the bird. The pact was made and Cycnus (now Cygnus), reclaimed his friend's body and gave him a proper burial. Touched by this show of unconditional love, Zeus placed his image in the sky. In Chinese mythology, Cygnus is known as the Magpie Bridge. Two lovers, one mortal and one a fairy girl, are separated by the sky river, the Milky Way. They reunite once a year with the help of magpies, who cluster together to form a celestial walkway.

VULPECULA

	FAMILY	HERCULES
	LOCATION	Fourth quadrant of the northern hemisphere
	BEST SEEN IN	September
	NAME MEANING	The Fox
	STARS	Anser (Alpha Vulpeculae), 23 Vulpeculae, 13 Vulpeculae, 31 Vulpeculae

This little fox may not at first glance light up the sky with its antics, but while there are no myths to speak of, this pale constellation does include the vibrant red star Anser. This star was once known as the goose (*anser* is the Latin for 'goose'), the catch of this wily fox. The two creatures were originally noted together by the 17th-century astronomer Johannes Hevelius, who called the constellation Little Fox with the Goose. He claimed the goose was a special delivery for Cerberus, the fearsome multi-headed dog who guarded the entrance to Hades, the Greek underworld. Cerberus once had a constellation of his own, but he was later absorbed into the constellation Hercules. The fox and the goose were eventually split into separate constellations, but more recently re-formed under the title Vulpecula.

HYDRA

⊙	**FAMILY**	HERCULES
⚡	**LOCATION**	Second quadrant of the southern hemisphere
👁	**BEST SEEN IN**	April
❓	**NAME MEANING**	The Female Water Snake
⭐	**STARS**	Alphard (Alpha Hydrae), Gamma Hydrae, Hydrobius (Zeta Hydrae), Beta Hydrae, HD 122430

The largest of all the constellations, tenacious Hydra is the giant female water serpent that the Greek hero Hercules killed as part of his twelve labours for King Eurytheus. According to the myth, Hydra had been running amok in the countryside, killing livestock and terrorising the people. Reputedly, the monster had nine heads (although accounts differ), only one of which was immortal. This is the head portrayed in the stars, which Hercules removed and buried beneath a boulder after a long and arduous battle with the serpent. In another myth, Hydra is unfairly blamed for delaying the crow belonging to Apollo on his quest to fetch the sun god a cup of water. Apollo quickly worked out that the crow was lying and, displeased with his cupbearer, the god cast the crow (Corvus), the cup (Crater) and Hydra high into the sky forever.

SEXTANS

◉	**FAMILY**	HERCULES
⚡	**LOCATION**	Second quadrant of the southern hemisphere
👁	**BEST SEEN IN**	April
❓	**NAME MEANING**	The Sextant
★	**STARS**	Alpha Sextantis, Beta Sextantis, Gamma Sextantis

Although it may be dim against the inky backdrop of the southern sky, Sextans represents what was for early astronomers a vital piece of equipment for measuring and mapping the night skies. It was originally named Sextans Uraniae, after the instrument used by 17th-century astronomer Johannes Hevelius, a large non-telescopic sextant 2m (6ft) tall with open sights that required naked-eye observations of the stars. First noted in 1687, this distant constellation reminds us of Hevelius's work and his preference for charting stars without the aid of a telescope. He must have had remarkable eyesight, as even his contemporary, the great English astronomer Edmond Halley (famed for his observations of Halley's comet), who had been sent by the Royal Society to persuade Hevelius to move with the times, was forced to admit that his bare-eye measurements were accurate, despite not using a telescope.

CRATER

◎	**FAMILY**	HERCULES
⚡	**LOCATION**	Second quadrant of the southern hemisphere
👁	**BEST SEEN IN**	April
❓	**NAME MEANING**	The Cup
★	**STARS**	Labrum (Delta Crateris), Alkes (Alpha Crateris), Gamma Crateris

This two-handed chalice belonged to Apollo. According to Greek legend, he sent his sacred bird, the crow Corvus, to collect some water in the cup for a magical ritual, but the crow was distracted from his errand by a tree of nearly ripe figs. Reluctant to miss such a great treat, Corvus decided to hang around for a few days until they were sweet enough to eat and he could gorge himself. When his belly was full, the cunning crow then filled the cup with water and returned, also carrying a water snake, with the excuse that the snake had blocked the stream's flow and delayed him. Wise Apollo saw through his lies and in a fit of pique cast the Crow (Corvus), the Cup and the Snake (Hydra) into the heavens. He cursed the crow for its deceit, scorching its feathers as black as coal and making it forever thirsty: hence the distinctive rasping caw it makes to this day.

CORVUS

◎	**FAMILY**	HERCULES
⚡	**LOCATION**	Third quadrant of the southern hemisphere
👁	**BEST SEEN IN**	May
❓	**NAME MEANING**	The Crow
★	**STARS**	Gienah (Gamma Corvi), Kraz (Beta Corvi), Algorab (Delta Corvi), Minkar (Epsilon Corvi)

The glossy black feathers of the crow weren't always so. His plumage was as white as snow, but various myths from around the world suggest they were tarnished, whether by stealing fire, flying too close to the sun, or angering the gods. The crow in this constellation is often associated with Apollo as his sacred bird. In one story he was given the task of guarding Apollo's lover Coronis. Though pregnant by the deity, her ardour had faded and she fell in love with a mortal man. On reporting her infidelity, the poor crow was blamed and its feathers scorched by Apollo, turning the bird pitch-black. The Babylonians identified Corvus with Adad, the god of rain and stormy weather. They noted that the constellation would appear just ahead of the rainy season: to them, a mystical and heavenly omen.

OPHIUCHUS

◉	**FAMILY**	HERCULES
⚡	**LOCATION**	Third quadrant of the southern hemisphere
◉	**BEST SEEN IN**	July
❓	**NAME MEANING**	The Serpent Bearer
★	**STARS**	Rasalhague (Alpha Ophiuchi), Sabik (Eta Ophiuchi), Zeta Ophiuchi

A man stands holding a giant serpent. He is Ophiuchus, the Serpent Bearer, commonly known as Asclepius, son of Apollo and the god of medicine. Asclepius was raised by the great teacher and healer, the centaur Chiron, and grew up to be gifted and wise in the healing arts. Legend has it that he took inspiration from the natural world, and that he once observed a snake bring a fellow serpent back to life by placing a certain herb on its head. Asclepius used the same herb to bring Glaucus, a prince of Crete, back from the dead, thus gaining the reputation that he had miraculous powers. Sometimes referred to as the thirteenth sign of the Zodiac, the constellation Ophiuchus sits on the imaginary circle, the celestial equator, that forms the sun's path through the sky in a dazzling array of bright stars.

SERPENS

⊙ **FAMILY**   HERCULES

● **LOCATION**   Third quadrant of the northern hemisphere

◉ **BEST SEEN IN**   July

? **NAME MEANING**   The Serpent

★ **STARS**   Unukalhai (Alpha Serpentis), **Eta Serpentis**,
                      Mu Serpentis, **Xi Serpentis**, **Gliese 710**,
                      Tau Serpentis (a star group consisting
                      of eight stars)

Enigmatic yet deadly, Serpens twines itself between the legs of Ophiuchus, who valiantly keeps the head of the serpent at bay with his left hand, while his right hand fends off the tail. Unusually, the constellation is made up of two separate star groups, consisting of the serpent's head, Serpens Caput, and the reptile's tail, Serpens Cauda. Unukalhai, the brightest star in the constellation, forms the neck of this magnificent creature while the tip of its tail reveals the star Alya. Serpens also has a more benign identity as the snake most commonly associated with the healer Asclepius, whose famous healing staff consisted of a rod entwined by a serpent. The snake, because it periodically sheds the outer layer of its skin to reveal fresh skin below, may have been seen as a symbol of rejuvenation by the ancient Greeks. The staff of Asclepius is still used as a symbol for many medical associations around the world today.

SCUTUM

FAMILY	HERCULES	
LOCATION	Fourth quadrant of the southern hemisphere	
BEST SEEN IN	August	
NAME MEANING	The Shield	
STARS	Ionnina (Alpha Scuti), Beta Scuti, R Scuti	

A tiny constellation, the fifth smallest in the night sky, Scutum was once known as Scutum Sobiescianum, meaning the 'Shield of Sobieski'. Named by the astronomer Johannes Hevelius in 1684, the constellation was a nod to King John III Sobieski of Poland, and an honour to mark his victorious Battle of Vienna. An ally of Leopold I of the Austrian Habsburg lands, King John came to Leopald's aid when the Ottoman Turks stormed the city of Vienna in 1683, successfully driving back the invading forces. Over time Sobiescianum was dropped, and this diminutive pattern of stars was simply Scutum. While it may be small, it contains one vibrant orange star named Ionnina (or Ioannina), which is about 21 times larger than the sun.

CENTAURUS

◉	**FAMILY**	HERCULES
⚡	**LOCATION**	Third quadrant of the southern hemisphere
◉	**BEST SEEN IN**	May
❓	**NAME MEANING**	The Centaur
★	**STARS**	Rigil Kentaurus (Alpha Centauri), **Hadar** (Beta Centauri), **Menkent** (Theta Centauri)

Half-man, half-horse, a hybrid creature of tremendous strength, the centaur finds its origins in the Babylonian Bison-man (or Bull-man) known as Kusarikku. Of similar stature, this creature had the body of a bison and the head of a man, although in some cases his human torso was attached. Associated with the Babylonian sun god, Shamash, he was known as the guardian of entrances. It's no surprise that this enormous constellation, the ninth largest in the sky, harbours two of the brightest stars in the heavens, Alpha Centauri (in *The Hitchhiker's Guide to the Galaxy*, the home of the Galactic Hyperspace Planning Council) and Beta Centauri, which make up the centaur's strident front legs. Some believe the image represents Chiron, a wise sage and centaur present in many Greek myths. He was famed for his knowledge of healing herbs and his surgical skills; he was also a renowned mentor to many notable figures in Greek mythology, including Apollo, Artemis and Asclepius, the god of medicine.

LUPUS

◉	**FAMILY**	HERCULES
⚡	**LOCATION**	Third quadrant of the southern hemisphere
◉	**BEST SEEN IN**	June
?	**NAME MEANING**	The Wolf
★	**STARS**	Alpha Lupi, Kekouan (Beta Lupi), Gamma Lupi

A beast once stalked the stars, a wild and nameless creature, impaled by the centaur (Centaurus) and raised high towards the constellation of Ara the Altar, as if presented in sacrifice. It had no name as such and was considered by the Greeks to be part of the Centaurus constellation until the second century BC when the ancient Greek astronomer Hipparchus declared it a separate constellation, naming it *Therium*, a feral animal. The ancient Romans called it *Bestia*, meaning 'beast', and it was only during the Renaissance that it acquired the name the Wolf (Lupus).

CORONA AUSTRALIS

⊙	**FAMILY**	HERCULES
⚡	**LOCATION**	Third quadrant of the southern hemisphere
👁	**BEST SEEN IN**	August
❓	**NAME MEANING**	The Southern Crown
★	**STARS**	Alphekka Meridiana (Alpha Coronae Australis), **Beta Coronae Australis, Gamma Coronae Australis, RX J1856.5-3754**

Dainty yet beautiful, this ornate tiara is recognised for its unique display of twinkling stars arranged in a horseshoe shape. On first glance, it looks like the crown falls at the feet of the Centaur, but most seem to think it sits on the head of Sagittarius, another distinguished centaur. To the ancient Greeks it was the wreath made of myrtle leaves that Dionysus, the god of wine and a son of Zeus, placed in the heavens as a memorial after he retrieved his mother, Semele, from the Underworld. It was an especially poignant occasion as the legend goes that Semele died in childbirth, and mother and child only met for the first time when Dionysus was fully grown and able to rescue her from the land of the dead. Although Semele was a mortal, a princess of Thebes, she was granted a place on Mount Olympus, the home of the gods, as a mother of a deity.

ARA

⊙	**FAMILY**	HERCULES
⚡	**LOCATION**	Third quadrant of the southern hemisphere
👁	**BEST SEEN IN**	July
❓	**NAME MEANING**	The Altar
⭐	**STARS**	Beta Arae, Choo (Alpha Arae), Zeta Arae

Dating back to the Babylonians, this sacred altar was thought to honour the mythical Tower of Babel, which also features in a story in the Old Testament. Set at a time when all humankind spoke one language, their ambitious aim to build a tower to the heavens displeased God, who punished their arrogance by causing the people to speak in many tongues. No longer in harmony, the people broke up into the tribes of the earth and the tower was abandoned. Though it is a dim and distant constellation, Ara plays an important role in Greek mythology. It was said to represent the altar of the Olympian gods, on which they swore allegiance to Zeus in the war against the Titans, the gods who ruled the heavens before them. Another version of this story suggests that Chiron the centaur (Centaurus) was the one to unite them by offering the Wolf (Lupus) as a magical sacrifice. Some accounts state that the smoke that billowed from the altar formed the Milky Way.

TRIANGULUM AUSTRALE

⊙	**FAMILY**	HERCULES
⚡	**LOCATION**	Third quadrant of the southern hemisphere
👁	**BEST SEEN IN**	July
❓	**NAME MEANING**	The Southern Triangle
⭐	**STARS**	**Atria** (Alpha Trianguli Australis), **Betria** (Beta Trianguli Australis), **Gatria** (Gamma Trianguli Australis)

The three brightest stars of this constellation form a perfect triangle, which the 17th-century German poet Philipp von Zesen (also known as Philippus Caesius) described as the 'Three Patriarchs', in honour of Abraham, Isaac and Jacob, the great patriarchs of the Israelites from the Old Testament. This tiny constellation, sixth smallest in the night sky, was first charted by the Dutch astronomer and cartographer Petrus Plancius on his celestial globe of 1589. Zesen claimed that its brightest star, an orange giant known as Atria, represented Abraham. This luminescent orb is 130 times bigger than the sun.

CRUX

⊙	**FAMILY**	HERCULES
⚡	**LOCATION**	Third quadrant of the southern hemisphere
👁	**BEST SEEN IN**	May
❓	**NAME MEANING**	The Southern Cross
⭐	**STARS**	**Acrux** (Alpha Crucis), **Mimosa** (Beta Crucis), **Gacrux** (Gamma Crucis), **Delta Crucis**

The smallest constellation in the night sky, what Crux lacks in size it makes up for in its colourful folklore. A Christian symbol of faith, the image of the cross is also found in many different cultures with differing yet meaningful symbols. The |Xam bushmen of Africa see Crux as three female lions gathered in waiting, while many Australian Aboriginal peoples know it is as the head of a larger constellation, the Emu in the Sky, which stretches across the Milky Way. They have given numerous interpretations to Crux itself, among others an eaglehawk, an eagle's footprint, a possum, a gum tree and a fish. In Brazil, Crux is a swarm of bees bursting from the beehive, while the Tainui Māori call it Te Punga, the trusty anchor of a great canoe that belonged to the hero Tama Rereti. Crux appears on the flags of Australia, New Zealand and Brazil.

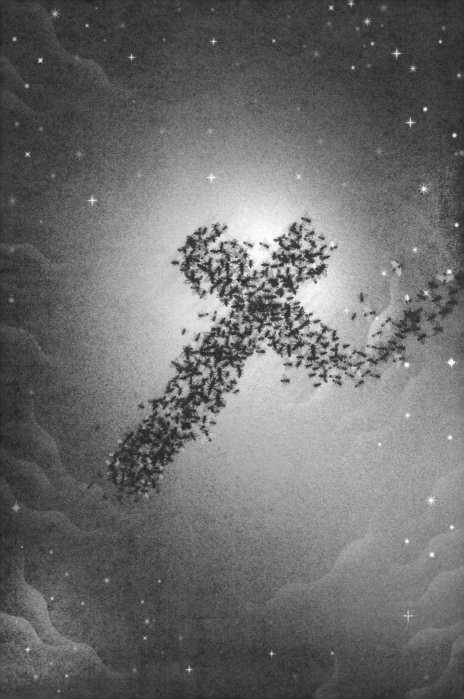

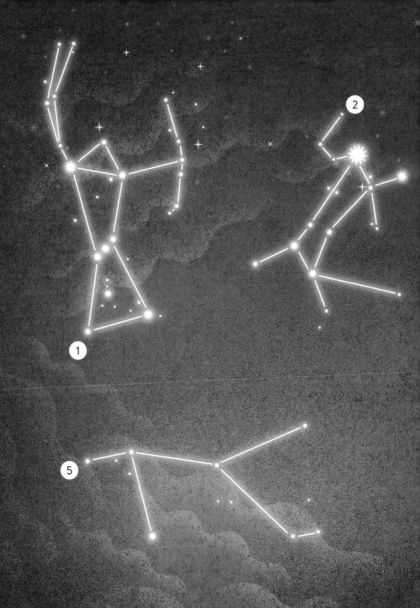

③

④

With the exception of the magical unicorn, each
constellation in this family is part of the same story:
an intrepid tale of the hunter and the hunted.
Moving like a flash through the night sky, each
character is intent on their mission: to catch or to
flee. Like so many constellations, the players in this
tale are frozen in time, caught for one single second
in the final leap, the bid for freedom as jaws nash
wildly just out of reach of their tails. It's an eternal
loop, and nobody knows how it will end. That is part
of Orion's charm. We, as the stargazers of these
constellations' future, get to decide their fate – and
if we don't like the ending, we can always change it
the next night.

O R I O N

ORION

◎	**FAMILY**	ORION
ⓘ	**LOCATION**	First quadrant of the northern hemisphere
◉	**BEST SEEN IN**	January
❓	**NAME MEANING**	The Hunter
⭐	**STARS**	**Rigel** (Beta Orionis), **Betelgeuse** (Alpha Orionis), **Bellatrix** (Gamma Orionis), **Alnilam** (Epsilon Orionis), **Alnitak** (Zeta Orionis), **Saiph** (Kappa Orionis), **Mintaka** (Delta Orionis)

A handsome hunter in Greek myth, Orion stands proudly in the heavens, with a shield and sword to hand and a belt around his waist. He faces the charge of the great bull Taurus, while his hunting dogs, Canis Major and Canis Minor, chomp at the feet of Lepus the hare. His belt was called Frigg's Distaff (a spindle used in spinning) in Norse mythology, and associated with the fertility goddess Frigg, who used her celestial spinning wheel to weave coloured clouds. The Aztecs called his belt the Fire Drill (a wooden hand drill used to start a fire). They timed their New Fire Ceremony, which occurred every 52 years, to coincide with Orion rising, a ritual to ensure the sun would shine for another 52-year cycle. The Native American Chinook tribe see Orion's belt and sword as two canoes in a race to catch the prize salmon that is the star Sirius from the Big River, the Milky Way.

CANIS MAJOR

◉	**FAMILY**	ORION
⚡	**LOCATION**	Second quadrant of the southern hemisphere
◉	**BEST SEEN IN**	February
❓	**NAME MEANING**	The Great Dog
★	**STARS**	Sirius (Alpha Canis Majoris), **Adhara** (Epsilon Canis Majoris), **Wezen** (Delta Canis Majoris)

Blazing from the head of Canis Major is Sirius, the brightest star in the night sky. Also known as the Dog Star, Sirius twinkles all the colours of the rainbow and was revered by the ancient Egyptians. They called it Sodpet, a goddess of fertility and prosperity; by Greaco-Roman times she had become associated with the goddess Isis. The first night of the year that Sodpet was seen marked the start of the Egyptian new year, and the yearly flooding of the Nile. Canis Major is often referred to as one of Orion's two hunting dogs, but may also represent the hound, Laelaps, who was so fast he always caught his prey, according to Greek myth. His master Cephalus set him to chase a troublesome beast known as the Teumessian fox (Canis Minor), but they were too well matched and the race continues to this day, with Canis Major in eternal pursuit of his prey through the stars.

CANIS MINOR

⊙	**FAMILY**	ORION
⚡	**LOCATION**	Second quadrant of the northern hemisphere
👁	**BEST SEEN IN**	March
❓	**NAME MEANING**	The Lesser Dog
★	**STARS**	Procyon (Alpha Canis Minoris), Gomeisa (Beta Canis Minoris), Gamma Canis Minoris

Sprinting like a tornado through the night sky, the smaller of Orion's two hunting dogs is often compared to the Teumessian fox of Greek myth. This wily creature was known for its speed and dexterity; it could not be outrun, not even when chased by the swift-running Laelaps (Canis Major). Some accounts declare that the Teumessian fox was a dreadful monster who roamed the land killing and devouring the people, and Cephalus and his hound Laelaps were sent to hunt him down. As neither beast could ever win the race, Zeus put an end to their antics, turning both of them to stone. To mark the end of the tale, he cast their images into the sky. Canis Minor's brightest star, Procyon, was known as 'before the dog' by the ancient Greeks, because it rises before the Dog Star, Sirius.

LEPUS

FAMILY   ORION

LOCATION   Second quadrant of the northern hemisphere

BEST SEEN IN   February

NAME MEANING   The Hare

STARS   Arneb (Alpha Leporis), Nihal (Beta Leporis),
                    Epsilon Leporis, Zeta Leporis

Described as being hunted by Orion and his dogs, the Hare is often depicted in antique star maps as lying at Orion's feet, or in mid-air as it leaps to freedom, seemingly frozen to the spot just as a live hare will do when sensing danger. Perhaps poor Lepus hopes that by staying completely still, he will escape the huntsman's keen eye, caught frozen for eternity. Although no myths are attached to this constellation, the hare was considered a sacred and powerful creature by the Celts (and sometimes a troublesome trickster too). Being most active at night, it was closely associated with the Moon. Ancients around the world have often linked the hare to the night sky, seeing its shape upon the moon's surface, or in the stars, as it appears in this constellation.

MONOCEROS

⊙	**FAMILY**	ORION
⚡	**LOCATION**	Second quadrant of the northern hemisphere
👁	**BEST SEEN IN**	February
❓	**NAME MEANING**	The Unicorn
★	**STARS**	Alpha Monocerotis, Gamma Monocerotis, Delta Monocerotis, Zeta Monocerotis, 13 Monocerotis, HD 48099, V838 Monocerotis

On noticing some space in the night sky between Orion and Hydra, astronomer Petrus Plancius decided to create the Unicorn constellation in 1612. Monoceros (the Latin for unicorn) finds its roots in myth. The ancient Greeks record that the *hippos monokeras* ('one-horned horse') was a pure-white horse with a purple-red head that sported a white and black horn, about a cubit long (roughly 50cm/20in), tipped with flaming red. They claimed that if you drank from a cup made from the horn it would cure convulsions and epilepsy and even provide an antidote to poison. Later seen as virtuous and pure, this horned horse became a Christian symbol of faith and love. While the stars are difficult to see, being faint against the backdrop of the sky, this beguiling creature continues to enchant, whether from the pages of a book or a smattering of stars in the heavens.

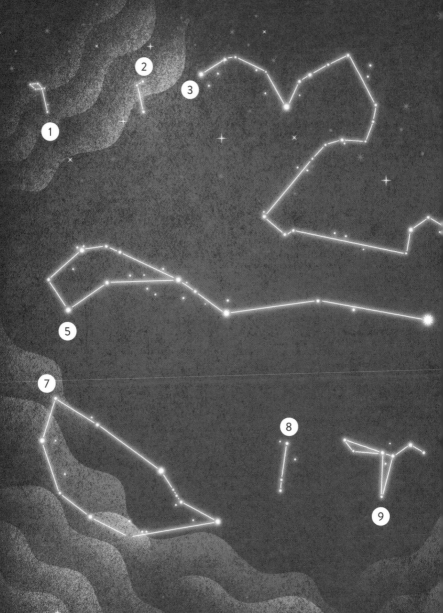

In certain lights, when gazing at the night sky it's easy to think of it as a giant ocean – a deep inky backdrop on which lights flicker, and soft waves flow steadily overhead. Stargazers of old thought the same thing, dividing the space to reveal the secrets of those rippling waters and the creatures that lived in its depths. This family of constellations is a reflection of that, a gathering of all things nautical. The mysteries of the deep can be seen in the names and the tales associated with the Heavenly Waters family of constellations. But while divers still plunder the ocean for treasure, all you have to do is gaze at the stars to experience that same feeling of wonder.

HEAVENLY WATERS

DELPHINUS

⊙	**FAMILY**	HEAVENLY WATERS
⚡	**LOCATION**	Fourth quadrant of the northern hemisphere
👁	**BEST SEEN IN**	September
❓	**NAME MEANING**	The Dolphin
⭐	**STARS**	Rotanev (Beta Delphini), Sualocin (Alpha Delphini), Gamma Delphini, Aldulfin (Epsilon Delphini)

Delphinus is the humble Dolphin, servant to Poseidon and friend to the beautiful sea nymphs the Nereids. In one Greek tale he is sent to retrieve the dark-eyed Amphitrite, a Nereid who had stolen Poseidon's heart. She didn't reciprocate his ardour, and had hidden away with her sisters far from the sea god's court. Poseidon couldn't have sent a better ambassador to press his suit for her hand than Delphinus. The dolphin's gentle and endearing ways persuaded Amphitrite to return to Poseidon's underwater palace, where she wed the love-struck god (one account records that the dolphin also conducted the wedding ceremony). In Roman myth, Delphinus is the dolphin that saves the famous poet and musician Arion from pirates and a watery doom. As a thank you, the god Apollo cast his image into the sky. The constellation includes the asterism Job's Coffin, so called because of its shape, although no one knows how it acquired this name.

EQUULEUS

⊙	**FAMILY**	HEAVENLY WATERS
⚡	**LOCATION**	Fourth quadrant of the northern hemisphere
👁	**BEST SEEN IN**	September
❓	**NAME MEANING**	The Little Horse
⭐	**STARS**	Kitalpha (Alpha Equulei), Pherasauval (Delta Equulei), Gamma Equulei

A tiny foal peeks from behind the mighty winged horse Pegasus, only her head in view. Some Greek myths claim that this is Celeris, either the offspring or brother of Pegasus. Others associate it with Hippe, the daughter of the centaur Chiron. When she found herself with child after being seduced by the nobleman Aeolus, Hippe fled to the mountains, afraid of what her father would do when he found out about her pregnancy. There she gave birth to a daughter, but living in fear that Chiron would discover her shame, she prayed to the gods for help, who obliged by changing her into a foal. Later Artemis, the goddess of both the hunt and of childbirth, hid her in the sky, where she remains, still obscured from her father's eyes.

ERIDANUS

⊙	**FAMILY**	HEAVENLY WATERS
⚡	**LOCATION**	First quadrant of the southern hemisphere
👁	**BEST SEEN IN**	December
?	**NAME MEANING**	The River
★	**STARS**	Achernar (Alpha Eridani), **Cursa** (Beta Eridani), **Zaurak** (Gamma Eridani), **Acamar** (Theta Eridani)

The name Eridanus is thought to have originated from the Star of Eridu, a Babylonian constellation named in honour of the city Eridu. This constellation is sacred to the deity Ea, who governed a region called Abzu, sometimes described as a freshwater ocean below the earth's surface. According to Greek legend, it is associated with the River Eridanus. Helios, the Titan sun god, offered to grant his son Phaeton one wish, but regretted his words when Phaeton asked to drive his father's sun chariot across the sky for one day. Helios reluctantly agreed, but as he feared, Phaeton was no match for the chariot's powerful horses and quickly strayed from the sun's prescribed path. Soon Phaeton was setting fire to the earth, creating vast deserts and raining chaos on mankind. Finally, Zeus cast a thunderbolt at the arrogant youth and the lad plunged to his doom into the Eridanus, a river later identified as the Po in northern Italy.

PISCIS AUSTRINUS

◉	**FAMILY**	HEAVENLY WATERS
⏀	**LOCATION**	Fourth quadrant of the southern hemisphere
◉	**BEST SEEN IN**	October
❓	**NAME MEANING**	The Southern Fish
★	**STARS**	Fomalhaut (Alpha Piscis Austrini), Epsilon Piscis Austrini, Delta Piscis Austrini

Numerous legends are associated with this constellation, the earliest being that of Atargatis, the Syrian goddess of fertility. Her temples in ancient times were said to include ponds filled with tame fish who liked to be petted. This bountiful deity fell into a lake one day and was rescued by a giant fish. To repay this good deed, she decreed that all those who ate fish would be punished. A later Greek version of the story names her Derceto, and recounts that she threw herself into the lake after having an affair with a handsome youth and bearing him a child. Instead of drowning, she was transformed into a beautiful mermaid. The ancient Greek astronomer and mathematician Eratosthenes proposed that the pair of bound fish that represent the constellation Pisces are the offspring of this vibrant constellation.

CARINA

⟳	**FAMILY**	HEAVENLY WATERS
⚡	**LOCATION**	Second quadrant of the southern hemisphere
👁	**BEST SEEN IN**	March
❓	**NAME MEANING**	The Keel
★	**STARS**	Canopus (Alpha Carinae), Miaplacidus (Beta Carinae), Avior (Epsilon Carinae), Eta Carinae

Outlining the keel of a ship, Carina once formed part of the much larger constellation Argo Navis, which was originally documented by the Greek astronomer Ptolemy in the 2nd century AD. It was thought by the ancient Greeks to represent the ship *Argo* that the hero Jason and the Argonauts sailed during his quest for the Golden Fleece. The Argo Navis constellation was so large and contained so many stars that the French astronomer Nicolas Louis de Lacaille divided it into three more manageable constellations in the 18th century: Carina, the keel; Puppis, the stern; and Vela, the sails. Carina contains the second-brightest star in the night sky after Sirius, the supergiant Canopus. This luminous beauty represents the tip of the blade from one of the ship's oars, and was used in ancient times as an important navigational star.

PUPPIS

⊙	**FAMILY**	HEAVENLY WATERS
⚡	**LOCATION**	Second quadrant of the southern hemisphere
👁	**BEST SEEN IN**	March
❓	**NAME MEANING**	The Stern
⭐	**STARS**	Naos (Zeta Puppis), Ahadi (Pi Puppis), Tureis (Rho Puppis)

The Stern is the largest of three constellations that once made up the ancient Greek constellation of *Argo* Navis. The vessel Argo was said to be the largest seen in the ancient world, and had rowing room for about 50 men (the exact number of Argonauts is not known, but most accounts agree that there were at least 50 crew). The ship was filled with many heroes to accompany Jason on his adventure, demigods of the ancient Greek world that included Hercules, Orpheus, the twins Pollux and Castor, Asclepius and Theseus. *Argo* was built by the shipwright Argos (or Argus), also one of the Argonauts, with magical assistance from Athena. She added a talking figurehead to the prow of the ship made from one of the sacred oak trees at the oracle of Dodona (dedicated to Zeus), which is said to have advised Jason during the course of his journey.

VELA

⊙	**FAMILY**	HEAVENLY WATERS
ⓘ	**LOCATION**	Second quadrant of the southern hemisphere
👁	**BEST SEEN IN**	March
❓	**NAME MEANING**	The Sails
★	**STARS**	Suhail al Muhlif/Regor (Gamma Velorum), Delta Velorum, Suhail (Lambda Velorum)

The third component of what was once the much larger constellation of Argo Navis, Vela represents the sails of the giant ship *Argo*. Some sources describe the boat as a type of warship or galley, large enough to carry Jason and the Argonauts from their home in the Greek region of Thessaly to the kingdom of Colchis on the Black Sea in search of the Golden Fleece. The Argonauts were blessed with the protection of both the goddess Hera and the goddess Athena in their venture. Athena helped in many ways with the project, including persuading the skilled sailor Tiphys to become the helmsman of the vessel. A Roman terracotta campana relief dating from the 1st century AD held in the British Museum collection depicts Tiphys holding a mast spar while a seated Athena helps adjust the sails.

PYXIS

◉	**FAMILY**	HEAVENLY WATERS
⚡	**LOCATION**	Second quadrant of the southern hemisphere
👁	**BEST SEEN IN**	March
❓	**NAME MEANING**	The Compass
★	**STARS**	Alpha Pyxidis, Beta Pyxidis, Gamma Pyxidis

Created by the astronomer Nicolas Louis de Lacaille in the 18th century, Pyxis was originally called *la Boussole* ('The Compass') and then given the Latin name *Pyxis Nautica* (The Mariner's Compass), but was eventually shortened to its current form. Thought to represent the magnetic compass used by seamen, this small constellation sits neatly in the area of the three constellations that replaced the now-obsolete Argo Navis. Because of this it's often called the fourth constellation of this group, even though it stands separately in the night sky.

COLUMBA

◉	**FAMILY**	HEAVENLY WATERS
⚡	**LOCATION**	First quadrant of the southern hemisphere
👁	**BEST SEEN IN**	February
❓	**NAME MEANING**	The Dove
★	**STARS**	Phact (Alpha Columbae), Wazn (Beta Columbae), Ghusn al Zaitun (Delta Columbae)

Recognised as a symbol of peace in many parts of the world, the graceful dove takes to the night sky in a flurry of luminous feathers. It was originally called *Columba Noachi*, meaning Noah's Dove, by the astronomer Petrus Plancius. According to the Biblical story of the Great Flood, after months afloat in the ark Noah released a dove and it returned with an olive leaf in its beak, a sign that the waters were receding. However, the dove could equally be the bird that Jason released at the entrance of the Symplegades, cliffs known as the Clashing Rocks that flanked the mouth of the Bosphorus. These terrifying rocks would move and crush anything that tried to pass between them. Jason's dove survived the ordeal thanks to its swift flight, although it did lose the tip of its tail. The Argonauts successfully followed suit, rowing for all they were worth, with only a damaged stern to mark the escapade.

3

We may look to the stars for inspiration, but the world around us also has the potential to stimulate our imagination. For Dutch astronomer and cartographer Petrus Plancius, this was the case. He paid close attention to his environment, and listened to the tales of those who'd travelled further afield. In the late 16th century he formed a close collaboration with the Dutch navigators Pieter Dirkszoon Keyser and Frederick de Houtman, and asked them to make close observations of the unmapped areas of the southern hemisphere during their travels to the East Indies in 1595 and 1596. The array of sights and sounds they reported to him triggered something in his scientific mind, and not only he could see the creatures described, he paid homage to them in the stars. This family of constellations conjures a colourful landscape that is both exotic and magical. To those who witnessed such wildlife first-hand, it must have seemed supernatural. Plancius understood this and captured these visions in the heavens on his celestial globe of 1598. The Johann Bayer Family is named after the German celestial cartographer Johann Bayer, famed for his star atlas, *Uranometria*, of 1603, which included beautiful engravings of the constellations.

JOHANN BAYER

HYDRUS

◉	**FAMILY**	JOHANN BAYER
⚡	**LOCATION**	First quadrant of the southern hemisphere
◉	**BEST SEEN IN**	December
❓	**NAME MEANING**	The Male Water Snake
★	**STARS**	Beta Hydri, Alpha Hydri, Gamma Hydri, HD 10180

The Latin for 'male water snake', Hydrus is separated from his much bigger cousin Hydra (the Female Water Snake) by the Milky Way. Hydrus is almost certainly named after the water snakes (or rather sea snakes) that 16th-century navigators Pieter Dirkszoon Keyser and Frederick de Houtman would have seen on their voyages at sea. These highly venomous snakes are related to the cobra and are found only in tropical waters. Perhaps Keyser and de Houtman encountered the yellow-bellied sea snake that lives far out at sea, and which can sometimes be seen in its thousands, gathered together on the surface of the water – a terrifying sight!

DORADO

⊙	**FAMILY**	JOHANN BAYER
⚡	**LOCATION**	First quadrant of the southern hemisphere
👁	**BEST SEEN IN**	January
❓	**NAME MEANING**	The Dolphinfish
⭐	**STARS**	Alpha Doradus, Beta Doradus, Gamma Doradus

While not a huge arrangement of stars, this modern constellation does contain most of the Large Magellanic Cloud, a small satellite galaxy that sits close to the Milky Way. Dorado was named after the exotic dolphinfish seen by Pieter Dirkszoon Keyser and Frederick de Houtman in the East Indies in the late 16th century. These eye-catching fish can grow to 2m (6½ft) in length and have striking blue, blue-green and yellow iridescent colouring. In Spanish, the dolphinfish is called a *dorado maverikos* (golden maverick), an appropriate moniker for a fish known for its bold, playful behaviour. At one time, the constellation was renamed Xiphias, the Swordfish, but the name reverted to Dorado.

VOLANS

◉	**FAMILY**	JOHANN BAYER
⚡	**LOCATION**	Second quadrant of the southern hemisphere
👁	**BEST SEEN IN**	March
❓	**NAME MEANING**	The Flying Fish
⭐	**STARS**	Beta Volantis, Gamma Volantis, Zeta Volantis

Bursting from the water, this tropical beauty glides through the air, translucent and shimmering. The constellation is named after the flying fish seen by Pieter Dirkszoon Keyser and Frederick de Houtman during their travels to the East Indies in the late 16th century. Although there are no myths attached to Volans, the first sight of a large school of these exotic, torpedo-shaped fish skimming across the water using their pectoral fins as 'wings' must have been a moment of wonder for the two explorers. The constellation was originally entitled Pisces Volans, meaning 'the flying fish', but during the mid-19th century this was shortened to Volans.

APUS

◎	**FAMILY**	JOHANN BAYER
⚡	**LOCATION**	Third quadrant of the southern hemisphere
👁	**BEST SEEN IN**	July
❓	**NAME MEANING**	The Bird of Paradise
★	**STARS**	Alpha Apodis, Gamma Apodis, Beta Apodis, Delta Apodis

The stunning bird of paradise sits like a tiny gem of colour in the night sky. Once believed to be a creature with no feet – Apus comes from the Greek *apous*, meaning 'footless' – this goes some way to explaining the name of this constellation. The first birds of paradise seen by Europeans were stuffed specimens brought back by sailors that consisted of just the head and body adorned with their fabulous plumage. As a result, zoologists at the time thought the bird lived its entire life on the wing and that it existed exclusively on a diet of heavenly dew. Interestingly the original name created by the astronomer Petrus Plancius caused much confusion in the early years. Plancius failed to use the services of a proofreader for his celestial globe of 1598, as the name appears there as *Paradysvogel Apis Indica*, which means 'Bird of Paradise Indian *Bee*'; he almost definitely intended it to read *Paradysvogel Avis Indica*, Bird of Paradise Indian *Bird*.

PAVO

⊙	**FAMILY**	JOHANN BAYER
⚡	**LOCATION**	Fourth quadrant of the southern hemisphere
👁	**BEST SEEN IN**	September
❓	**NAME MEANING**	The Peacock
★	**STARS**	Peacock (Alpha Pavonis), **Beta Pavonis**, Delta Pavonis

The glorious peacock with its all-seeing eyes is a resplendent sight. In the night sky it watches over us, a symbol of beauty, integrity and protection. While Pavo is most likely named after the Java Green Peacock spotted by the 16th-century Dutch navigators Pieter Dirkszoon Keyser and Frederick de Houtman on their travels, the peacock is the sacred bird of Hera. According to Greek myth, Zeus fell in love with the nymph Io, but when his wife, Hera, became suspicious he turned the girl into a white heifer. Quick-witted Hera, sensing a trick, asked for the heifer as a gift, which Zeus could hardly refuse. Hera employed the hundred-eyed giant Argus to watch over Io, the perfect guard. However, Zeus had other ideas. He sent Hermes, the messenger of the gods, to kill Argus and free the nymph. Hera, feeling pity for the giant's demise, took his eyes and scattered them on the peacock's tail feathers, where they remain to this day.

GRUS

⊙ **FAMILY** — JOHANN BAYER

⚡ **LOCATION** — Fourth quadrant of the southern hemisphere

👁 **BEST SEEN IN** — October

❓ **NAME MEANING** — The Crane

⭐ **STARS** — Alnair (Alpha Gruis), Tiaki (Beta Gruis),
Aldhanab (Gamma Gruis)

Astronomers of the Elizabethan era all agreed that this constellation
represented a long-necked wading bird but not what type, so it was
a contest held among the crane, the heron and the flamingo. The
elegant Grus (the Latin word for 'crane') won the prize, and was almost
definitely named after the cranes that Pieter Dirkszoon Keyser and
Frederick de Houtman would have seen on their travels in the 16th
century. It is worth noting that the crane was also the sacred bird of
Hermes, who was the herald and trusted messenger of Zeus. Ancient
accounts claim that some or all of the letters of the Greek alphabet
were based on the various patterns a flock (or dance) of cranes
performs in flight – and even that it was Hermes who spotted these
flight shapes and turned them into letters. Grus' most vibrant star is
the blue-white giant Alnair, a fitting name – in Arabic, *Al Na'ir* means
'the bright one' – as it is about 23 times brighter than our sun.

PHOENIX

⊙	**FAMILY**	JOHANN BAYER
⬧	**LOCATION**	First quadrant of the southern hemisphere
👁	**BEST SEEN IN**	November
❓	**NAME MEANING**	The Phoenix
★	**STARS**	Ankaa (Alpha Phoenicis), Beta Phoenicis, Gamma Phoenicis

With its startling flame-like hues, the mythical Phoenix lights up the night sky. A symbol of rebirth, according to Greek legend this sacred bird could live for 500 years and then be restored to new life. There are several versions of the regeneration myth, the most striking accounts describe how at the end of its long life, the bird would make a funeral pyre consisting of spices and herbs. Beseeching the gods to send down a heavenly flame to set the pyre alight, the phoenix would be consumed by the flames, only to rise again from the ashes, newly reborn in its youthful form. To give its former self a sacred burial, the young bird would then place the ashy remains in a hollowed-out egg made of myrrh and fly with it to the ancient Egyptian city of Heliopolis to lay it down in the temple of the sun god Ra.

TUCANA

⊙	**FAMILY**	JOHANN BAYER
⚡	**LOCATION**	First quadrant of the southern hemisphere
👁	**BEST SEEN IN**	November
❓	**NAME MEANING**	The Toucan
⭐	**STARS**	Alpha Tucanae, Gamma Tucanae, Zeta Tucanae

With an enormous canoe-shaped bill, which is often brightly coloured and with distinctive markings, the toucan is a striking sight. A South American forest bird, it was given a place in the night sky by astronomer Petrus Plancius in his celestial globe of 1598. Since then it has been likened to the hornbill, an exotic bird found in the East Indies, but Plancius's version held firm and the toucan still has a presence in the starry heavens.

INDUS

⊙ **FAMILY**   JOHANN BAYER

⚡ **LOCATION**   Fourth quadrant of the southern hemisphere

◉ **BEST SEEN IN** September

❓ **NAME MEANING** The Indian

★ **STARS**   The Persian (Alpha Indi), Beta Indi, Epsilon Indi

The Indian stands in deep concentration, an arrow or slender spear raised in one hand and several more tucked under his other arm. He is primed for the celestial hunt. Though it is hard to tell where he hails from – some believe he is Madagascan, while others say he could be from the East Indies or even the Americas – and he represents no legendary figure, Indus stands for all indigenous peoples, he is Everyman in the throes of survival, captured in the stars.

CHAMAELEON

⊙	**FAMILY**	JOHANN BAYER
⚡	**LOCATION**	Second quadrant of the southern hemisphere
⊙	**BEST SEEN IN**	April
❓	**NAME MEANING**	The Chameleon
★	**STARS**	Alpha Chamaeleontis, Gamma Chamaeleontis, Beta Chamaeleontis

Don't misjudge this small constellation – a tiny lizard it may be, but the chameleon in real life has the extraordinary ability to change the colour and patterns of its skin to blend in with its environment. The chameleon also changes colour depending on its mood and as a form of communication. Pieter Dirkszoon Keyser and Frederick de Houtman would have almost definitely encountered this intriguing reptile during their stay in Madagascar, as this large tropical island is host to 59 species of chameleon. In his time, the 16th-century Dutch cartographer and engraver Jodocus Hondius was renowned for his beautifully executed maps and globes. His witty rendering of the lively lizard on one of his celestial globes shows the reptile in the act of catching the Fly (Musca) with its long, outstretched tongue. One of Hondius's celestial globes also appears in the painting *The Astronomer* by the great Dutch artist Johannes Vermeer.

MUSCA

FAMILY JOHANN BAYER

LOCATION Third quadrant of the southern hemisphere

BEST SEEN IN May

NAME MEANING The Fly

STARS Alpha Muscae, Beta Muscae, Delta Muscaey

When Petrus Plancius first charted this small constellation in 1598, it was nameless. The astronomer Johann Bayer gave it the name *Apis*, meaning 'the bee', and while this stuck for a while, eventually the Dutch cartographer Willem Janszoon Blaeu gave it the name Musca (the Latin word for 'insect') in 1602. Also known as the Southern Fly, over time the location was dropped and the ever-persistent Musca remained hovering in the night sky.

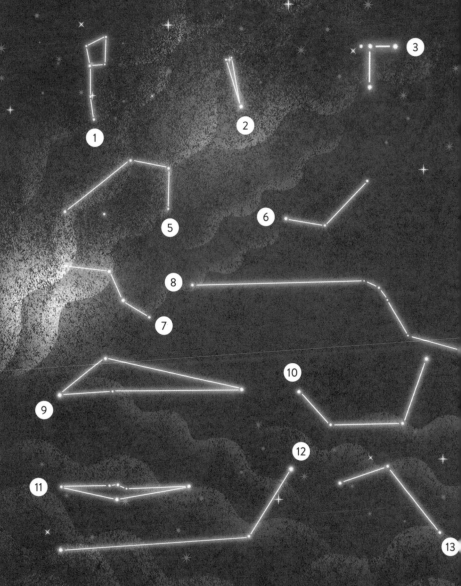

While there's an element of storytelling in most of the constellation families, the Lacaille family is an exception. It was created by the 18th-century French astronomer Nicolas Louis de Lacaille during his expedition to the Cape of Good Hope 1751–4 to observe the stars of the southern sky. These groups of stars are practical in nature, used to commemorate the tools and instruments used in navigation, science and the arts. Some might suggest this lacks imagination, but the opposite is true. Lacaille wished to honour the items that were instrumental in the birth of the Enlightenment, also referred to as the Age of Reason, in the 18th century. This influential period placed great importance on rational thought, human equality and freedom. The Enlightenment encouraged independent thinking, objectivity, self-reliance and profound questioning – qualities that would drive new discoveries in the sciences and inform intellectual thought, as well as see the emergence of the neoclassical style in the arts. Lacaille mixed the ethereal grace of the faintest stars with the effective power of these implements to create a toolkit filled with stardust, and although they might not be influenced by the gods, they've proved essential in helping mankind move forwards, and ultimately reach for the stars.

LACAILLE

NORMA

⊙	**FAMILY**	LACAILLE
⚡	**LOCATION**	Third quadrant of the southern hemisphere
👁	**BEST SEEN IN**	July
❓	**NAME MEANING**	The Carpenter's Square
⭐	**STARS**	Gamma-2 Normae, Epsilon Normae, Iota-1 Normae

This faint constellation represents a carpenter's square, an L-shaped measuring tool used for defining right angles. Lacaille originally called it the *l'Equerre et la Regle*, referring to a draughtsman's set square and ruler, and he placed this tiny arrangement of stars to sit neatly between Lupus and Ara. A tool in use since ancient times – both the Roman architect Vitruvius and the Roman scholar Pliny mention the instrument in their writings – Norma may be a nod to the common thread that connects 18th-century craftsmen to their classical forebears and the influence of the classical world on furniture design and woodwork during the Enlightenment. Lacaille later gave the constellation the Latin name Norma (the Carpenter's Square), although the ruler still appears in the drawing.

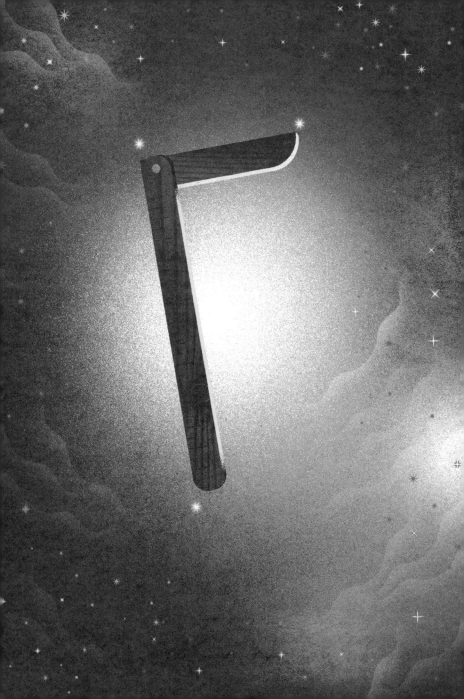

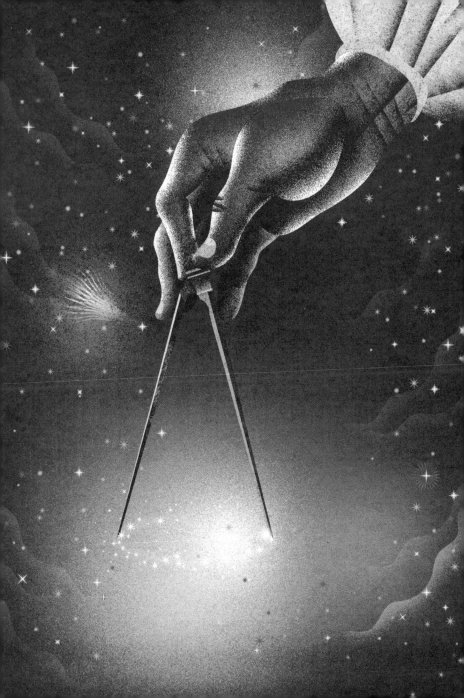

CIRCINUS

⊙	**FAMILY**	LACAILLE
⚡	**LOCATION**	Third quadrant of the southern hemisphere
👁	**BEST SEEN IN**	June
❓	**NAME MEANING**	The Compass
⭐	**STARS**	Alpha Circini, Beta Circini, Gamma Circini

The fourth smallest constellation in the night sky, Circinus (the Latin word for 'compass') refers to the drawing compass used to create perfect circles. However, on old star charts it is depicted as a pair of dividers used for measuring the distance between two points, used by draughtsmen and for reading maps. In the hands of a skilled architect, the humble drawing compass produced one of the key building blocks of architectural design in the 18th century: the circle, or sphere. This, together with the square, or cube, provided the basis for a system of proportions that took its inspiration from the work of 16th-century Italian architect Andrea Palladio. He in turn was heavily influenced by the writings of the Roman architect Vitruvius. In the 18th century the results were the many elegant buildings inspired by classical antiquity for which the period is famed.

TELESCOPIUM

◎	**FAMILY**	LACAILLE
⚡	**LOCATION**	Fourth quadrant of the southern hemisphere
👁	**BEST SEEN IN**	August
❓	**NAME MEANING**	Fourth quadrant of the southern hemisphere
★	**STARS**	Alpha Telescopii, Zeta Telescopii, Epsilon Telescopii

South of the constellations Sagittarius and Corona Australis, Telescopium peers from the night sky. It is depicted on Lacaille's star map of 1756 as a form of refractor telescope, possibly of the type known as an aerial telescope that was used at the time at the Paris Observatory. Once known as *Tubus Astronomicus* (Astronomical Tube), this faint patination of stars was squeezed uncomfortably amid its neighbouroughing constellations; Telescopium's boundaries were only finally defined in 1930 by Belgian astronomer Eugène Delporte. Remarkably, Lacaille himself used only a small 66-cm (26-inch) refracting telescope with a 1.3-cm (½-inch) aperture to catalogue and map the southern skies. From his observation point in Cape Town near the shore of Table Bay, he spent night after night peering at the southern stars, apparently kept company only by his small dog Grisgris. During his two-year stay there it is believed he observed in the region of 10,000 stars.

MICROSCOPIUM

◎	**FAMILY**	LACAILLE
⚡	**LOCATION**	Fourth quadrant of the southern hemisphere
👁	**BEST SEEN IN**	September
❓	**NAME MEANING**	The Microscope
★	**STARS**	Gamma Microscopii, Epsilon Microscopii, Theta Microscopii

Described by Lacaille as a 'pipe above a square box', Microscopium is a small, faint constellation that is hard to pinpoint from northern locations. Named after a type of microscope that uses more than one lens, this arrangement of stars lies to the south of Capricornus. The Dutch spectacle-maker Zacharias Janssen is generally credited with inventing the microscope in the 1590s. In the following century, the Dutch scientist Antonie van Leeuwenhoek, widely known as the father of microbiology, created more powerful optical lenses that enabled him to observe single-celled organisms, notably bacteria, for the first time – a revolutionary discovery. He was the first scientist to examine rainwater, and marvelled to see the 'exceeding living animals' not visible to the naked eye. His research helped pave the way to the science of microbiology, which saw significant advances in the 19th century.

SCULPTOR

◎	**FAMILY**	LACAILLE
⚡	**LOCATION**	First quadrant of the southern hemisphere
◉	**BEST SEEN IN**	November
❓	**NAME MEANING**	The Sculptor
★	**STARS**	Alpha Sculptoris, Beta Sculptoris, Gamma Sculptoris

A carved head sits neatly on a table; two chisels and a mallet lie beside it as if the artist has just left the room. This image is how the constellation was first depicted, when it was originally known as *l'Atelier du Sculpteur*, 'the sculptor's studio'. The neoclassical style informed much of the artistic production, including sculpture from the mid-18th century onwards. The movement was inspired by archaeological finds of ancient Roman artefacts that were viewed as ideal representations of beauty. Artists turned away from the excesses of the baroque and rococo periods to embrace the grace and serenity of these much-admired classical works, and to find a purer, nobler expression. The French sculptor Jean-Baptiste Pigalle (1714–1785) was a popular exponent of a style that married elements of the rococo and classicism, and produced many statues drawing on mythological figures, including his memorable *Mercury Attaching His Wings* (1744). Lacaille later gave the constellation the Latin name *Appartus Sculptoris*, but over the years it has evolved into one word.

FORNAX

FAMILY   LACAILLE

LOCATION   First quadrant of the southern hemisphere

BEST SEEN IN December

NAME MEANING The Furnace

STARS   Dalim (Alpha Fornacis), Beta Fornacis, Nu Fornacis

Lacaille's Latin name for the Furnace was *Fornax Chimiae* (the Chemical Furnace). On Lacaille's star chart of 1756, it is depicted as a small furnace with distillation apparatus. This would have been an essential piece of equipment for chemists at the time, who were experimenting with separating substances through the technique of distillation – i.e. by the evaporation and condensation of liquids – to reveal new components. While the greatest leaps forward in chemistry using this technique would be in the 19th century, one unexpected boon to the artistic world was the accidental discovery of the synthetic pigment Prussian blue by the German pigment-maker Heinrich Diesbach in around 1704. Prior to this, the only deep-blue pigment available to artists was one made from the semi-precious stone lapis lazuli, which was imported all the way from Afghanistan and was fabulously expensive. The chemical element of the constellation's name was eventually dropped and the constellation became simply Fornax.

CAELUM

⊙	**FAMILY**	LACAILLE
⚡	**LOCATION**	First quadrant of the southern hemisphere
👁	**BEST SEEN IN**	January
❓	**NAME MEANING**	The Engraving Tool
★	**STARS**	Alpha Caeli, Gamma Caeli, Beta Caeli

The great printmaker in the heavens, who scores the pattern of the stars in the night sky, works with a tool like this. This diminutive constellation forms yet another part of the astronomer's toolkit laid out in the stars, and was originally named *les Burins* on Lacaille's celestial map of 1756. It depicted fine chisels made of steel (burins) used for engraving on metal plates, tied together by ribbon. Thanks to copperplate engraving and printing, astronomers of the time could find a wider audience for their celestial maps, whether produced as single sheets or bound in atlases, which were highly sought after for both their educational value and their beauty. One of the most prized was Johann Bayer's atlas *Uranmetria*, published in 1603, with exquisite engravings by the German artist Alexander Mair. Lacaille later gave the constellation the Latin name *Caelum Scalptorium* (the Engraver's Chisel), which was eventually contracted to Caelum.

HOROLOGIUM

⊙ **FAMILY** LACAILLE

🜨 **LOCATION** First quadrant of the southern hemisphere

👁 **BEST SEEN IN** December

❓ **NAME MEANING** The Clock

★ **STARS** Alpha Horologii, R Horologii, Beta Horologii

While it may be small and faint to the eye, this constellation harbours an enormous supercluster containing 5,000 galaxy groups known as the Horologium Supercluster. Lacaille gave this group of stars the Latin name Horologium (the Clock). It's thought that Lacaille named the constellation after the clock he used to time his astronomical observations. He may also have had in mind the 17th-century Dutch scientist Christiaan Huygens, who designed the first pendulum clock in 1656 (Lacaille's star map shows a pendulum clock), which was a significant breakthrough in the history of clockmaking that allowed for a far more accurate measurement of time.

OCTANS

⊙	**FAMILY**	LACAILLE
⚡	**LOCATION**	Fourth quadrant of the southern hemisphere
👁	**BEST SEEN IN**	October
❓	**NAME MEANING**	The Octant
★	**STARS**	Nu Octantis, Beta Octantis, Delta Octantis, Polaris Australis

Circumpolar to the South Pole, this extremely faint constellation never sets below the horizon. Lacaille first named it *l'Octans de Reflexion*, meaning the 'reflecting octant', which later evolved into the Octans. An octant is a portable navigational tool consisting of an arc at the base measuring one-eighth of a circle (hence octans, referring to an eighth), a small sighting telescope and a set of tiny mirrors. The octant enabled navigators to read the angle of stars and planets in relation to the horizon, information that was used in conjunction with a nautical almanac, a yearly publication of astronomical information, to establish their location. The constellation was sometimes referred to as the Octans Hadleianus, in honour of John Hadley, the 18th-century English mathematician who invented the measuring instrument.

MENSA

⊙	**FAMILY**	LACAILLE
⚡	**LOCATION**	First quadrant of the southern hemisphere
👁	**BEST SEEN IN**	January
❓	**NAME MEANING**	The Table Mountain
★	**STARS**	Alpha Mensae, Gamma Mensae, Beta Mensae

Lacaille's Latin name for this small constellation was *Mon Mensae* (Table Mountain), but it was later contracted to Mensa (the Latin word for 'table'). He named this arrangement of stars after the iconic mountain of the same name that overlooks Cape Town in South Africa, a flat-topped mountain that is often covered in cloud. Cloaked in the ethereal Large Magellanic Cloud (a galaxy close to the Milky Way), the constellation too has the appearance of a mist-shrouded mountaintop. Mensa is the most southerly group of stars and cannot be seen from the northern hemisphere.

RETICULUM

◎	**FAMILY**	LACAILLE
⚡	**LOCATION**	First quadrant of the southern hemisphere
👁	**BEST SEEN IN**	January
❓	**NAME MEANING**	The Reticle
★	**STARS**	Alpha Reticuli, Beta Reticuli, Epsilon Reticuli

The first constellation created in this area of the night sky was not by Lacaille but was the work of the German astronomer Isaac Habrecht II. It appeared on Habrecht's celestial globe of 1621 as the constellation Rhombus. Lacaille changed the name in the 18th century to *le Réticule Rhomboide* ('the rhomboid reticle') to commemorate the tiny rhomboidal-shaped reticle – a focusing point made up of fine lines – in the centre of his telescope's eyepiece. This would have allowed him, for example, to keep a given star in the centre of his field, and to make precise measurements. Without the aid of this feature, he would not have been able to observe the night sky so closely. Lacaille later gave the constellation the Latin name Reticulus (related to the Latin reticulum meaning 'small net') on his star map of 1763, which was changed to Reticulum in the 19th century.

Auriga

Delphinus

Serpens

PICTOR

⊙	**FAMILY**	LACAILLE
⚡	**LOCATION**	First quadrant of the southern hemisphere
👁	**BEST SEEN IN**	February
❓	**NAME MEANING**	The Painter's Easel
⭐	**STARS**	Alpha Pictoris, Beta Pictoris, Gamma Pictoris

Thought to represent the painter's easel, delicate Pictor was introduced by Lacaille in 1756 as *le Chevalet et la Palette*, meaning 'the easel and the palette'. Lacaille did not live to see the full flowering of art in the Enlightenment, which found its greatest expression in the late 18th century with the neoclassical movement and the works of the French artist Jacques-Louis David. In the first half of the century the rococo style was *de rigueur*, particularly in France, with its whimsical, lighthearted scenes of predominantly pastoral idylls. Neoclassical painting, on the other hand, was heavily influenced by classical antiquity and the high ideals of the Enlightenment. The focus turned to the treatment of noble subjects such as stoicism and self-sacrifice, a message that was reinforced by the use of straight lines, clearly delineated forms and subdued colours. Lacaille subsequently named the constellation *Equuleus Pictorius* (the Painter's Easel) in his 1763 star chart, which was later shortened to Pictor.

ANTLIA

◉	**FAMILY**	LACAILLE
⚡	**LOCATION**	Second quadrant of the southern hemisphere
◉	**BEST SEEN IN**	April
❓	**NAME MEANING**	The Air Pump
★	**STARS**	Alpha Antliae, Epsilon Antliae, Iota Antilae

Antlia (the Latin word for 'pump') was created to commemorate the air pump, a single-cylinder pump used in experiments during the 1670s. Lacaille gave it the Latin name *Antlia Pneumatica* (the Air Pump) on his star map of 1763, and it was later shortened to Antlia. The original pump was invented by the 17th-century French physicist Denis Papin, who later went on to design a type of early pressure cooker known as the steam digester. He is also known for his pioneering research into harnessing steam pressure in a piston-cylinder system, which anticipated the development of steam engines. The first steam-powered machines were developed by the Spanish inventor Jerónimo de Ayanz and the English engineer Thomas Savery to address the problem of pumping floodwater out of deep coal mines, but they had their flaws. The first effective steam engine was developed by the English engineer Thomas Newcomen in 1712, commissioned by a Staffordshire coal mine.

Monthly Guide to
the Best Time to See
CONSTELLATIONS

The stars you can see in the night sky at any given time are dependent on where you live. To check if and when you can see a constellation from your location, refer to a local astronomy website or a stargazing guidebook specific to your region. Most constellations are visible for several months, rising and setting over a period of time. The monthly guide below gives the optimal month to see a constellation at around 9.00pm in the evening (a popular time for viewing the stars), and applies to both the northern and southern hemispheres. Included are the latitudes from which each constellation should be visible for the UK, Eire, the USA, Canada, South Africa, Australia and New Zealand.

A QUICK GUIDE TO LATITUDES

Latitudes are a series of imaginary parallel lines that run east to west around the globe, used to calculate the distance from the equator to the North and South Poles. Lines of latitude are indicated by degrees, with 0 degrees at the equator, +90° (90 degrees north) at the North Pole and -90° (90 degrees south) at the South Pole.

Compare the latitudes of the countries detailed below with the latitudes between which the constellations are visible to find out whether you can see a given constellation from your location (latitudes are rounded to the nearest whole number).

UK lies between +61° and +50° *(from the Shetland Islands in the north to Land's End, Cornwall, in the south)*, with **BELFAST** lying at +55°, **EDINBURGH** lying at +56°, **LONDON** lying at +52°. **EIRE** lies between +55° and +52°, with **DUBLIN** lying at +53°. **USA** the 48 contiguous states of the USA lie between +50° and +25°; **WASHINGTON DC** lies at +39°; **ALASKA** lies between +71° to +53°, Juneau at +58°; **HAWAII** lies between +22° *(Kauai)* to +19°, **HONOLULU** lies at +21°.

CANADA lies between +42° and +81° *(Ellesmere Island)*; OTTAWA lies at +45°. SOUTH AFRICA lies between -25° and -35°; PRETORIA lies at -26°. AUSTRALIA lies between -12° and -41° *(Tasmania)*, CANBERRA lies at -35°. NEW ZEALAND lies between -35° and -47° (Stewart Island); WELLINGTON lies at -41°.

JANUARY
Caelum +40° to -90°
Dorado +15° to -90°
Mensa +0° to -90°
Orion +85° to -75°
Reticulum +20° to -90°
Taurus +90° to -65°

FEBRUARY
Auriga +90 to -40°
Camelopardalis +90° to -10°
Canis Major +60° to -90°
Columba +45° to -90°
Gemini +90° to -60°
Lepus +60° to -90°
Monoceros +75° to -85°
Pictor +25° to -90°

MARCH
Cancer +90° to -60°
Canis Minor +85° to -75°
Carina +20° to -90°
Lynx +90° to -35°
Puppis +40° to -90°
Pyxis +50° to -90°
Vela +30° to -90°
Volans +10° to -90°

APRIL
Antlia +45° to -90°
Chamaeleon +0° to -90°
Crater +65° to -90°
Hydra +60° to -90°
Leo +90° to -65°
Leo Minor +90° to -45°
Sextans +80° to -80°
Ursa Major +90° to -30°

MAY
Canes Venatici +90° to -40°
Centaurus +30° to -90°
Coma Berenices +90° to -60°
Corvus +60° to -90°
Crux +20° to -90°
Musca +10° to -90°
Virgo +80° to -80°

JUNE
Boötes +90° to -50°
Circinus +20° to -90°
Libra +65° to -90°
Lupus +35° to -90°
Ursa Minor +90° to -10°

JULY
Apus +5° to -90°
Ara +25° to -90°
Corona Borealis +90° to -50°
Draco +90° to -15°
Hercules +90° to -50°
Norma +30° to -90°
Ophiuchus +80° to -80°
Scorpius +40° to -90°
Serpens +80° to -80°
Triangulum Australe
  +15° to -90°

AUGUST
Corona Australis
  +40° to -90°
Lyra +90° to -40°
Sagittarius +55° to -90°
Scutum +70° to -90°
Telescopium +30° to -90°

SEPTEMBER
Aquila +85 to -75°
Capricornus +60° to -90°
Cygnus +90° to -40°
Delphinus +90° to -70°
Equuleus +90° to -70°
Indus +25° to -90°
Miscroscopium +45° to -90°
Pavo +15° to -90°
Sagitta +90° to -70°
Vulpecula +90° to -55°

OCTOBER
Aquarius +65° to -90°
Cepheus +90° to -10°
Grus +35° to -90°
Lacerta +90° to -35°
Octans +5° to -90°
Pegasus +90° to -60°
Pisces Austrinus
  +50° to -90°

NOVEMBER
Andromeda +90° to -40°
Cassiopeia +90° to -20°
Phoenix +30° to -90°
Pisces +90° to -65°
Sculptor +50° to -90°
Tucana +15° to -90°

DECEMBER
Aries +90° to -60°
Cetus +70° to -90°
Eridanus +60° to -90°
Fornax +50° to -90°
Horologium +20° to -90°
Hydrus +5° to -90°
Perseus +90° to -35°
Triangulum +90° to -50°

Gods, Goddesses, Nymphs, Nobles, and Divine Heroes and Heroines

Note: the Greek gods are referred to as Olympian gods as they lived on Mount Olympus

AEGIPAN a Greek rustic faun, possibly an alter ego for the Olympian god Pan

AMPHITRITE a Greek sea nymph known as a Nereid, queen of the sea, wife to Poseidon

ANDROMEDA a princess of Ethiopia, married to Perseus

APHRODITE the Olympian goddess of love, beauty and fertility, wife of Hephaestus mother to Eros

APOLLO the Olympian god of the sun, also known as the god of light, knowledge, truth, prophecy and oracles, poetry, music, art, archery, and god of healing and of disease

ARES the Olympian god of war

ARIANRHOD the Celtic goddess of the moon and stars

ARTEMIS the Olympian goddess of the hunt, wild animals, wilderness and childbirth; protectress of girls and maidens

ASCLEPIUS the Olympian god of medicine, son of Apollo and the mortal Coronis

ATARGATIS the Syrian goddess of fertility, known to the Greeks as Derceto

ATHENA the Olympian goddess of warfare, wisdom and divine guidance, invention, crafts, and patroness of heroic endeavours, the state and all social institutions, including law and order

BISON-MAN (*Bull-man*) the Babylonian guardian of entrances; also known as Kusarikku

BOÖTES the archetypal herdsman, identified with the demi-god Philomelus

CASTOR twin of Pollux, joint gods of St Elmo's Fire (a faint light caused by an electrical discharge occurring on pointed objects such as ship's mast), and horsemanship; patron gods of sailors, guests and travellers, and the Olympic Games

CHIRON a noble centaur, immortal son of the Titan god Cronus and the sea nymph Philyra

CRONUS the king of the Titans, a race of giant gods who ruled before the Olympians, husband to Rhea

CUPID the Roman god of love, desire and attraction, son of Venus

DEMETER, the Olympian goddess of agriculture, fertility and the harvest

DERCETO the Greek name for the Syrian goddess Atargatis

DIKE the Olympian goddess of justice and righteousness

DIONYSUS the Olympian god of wine and winemaking, merriment and wild frenzy, son of Zeus and the mortal Semele

EA the Babylonian god of water

ERICHTHONIUS a legendary Greek hero and an early king of Athens

EROS the Olympian god of love and desire, son of Aprhodite

FRIGG the Norse goddess of fertility and marriage

GAIA, the Greek primordial mother goddess of the earth

HADES the Olympian king of the underworld and god of the dead

HAPI the Egyptian god of the Nile

HELIOS the Titan god of the sun, father to Phaeton

HEPHAESTUS the Olympian god of blacksmiths and fire, husband to Aphrodite

HERA the queen of the Olympian gods, the goddess of marriage and protectress of women and childbirth, wife to Zeus

HERACLES a divine Greek hero, known to the Romans as Hercules, son of Zeus and the mortal Alcemene

HERCULES a divine Roman hero, known to the Greeks as Heracles, son of Jupiter (Zeus) and the mortal Alcemene

HERMES the herald and messenger of the Olympian gods, also known as the god of communication and diplomacy, commerce and travellers, as well as of tricksters and thieves; protector of flocks, and the divine guide who led the dead to the underworld

HESPERIDES Greek nymphs, daughters of evening

HORAE the four Olympian goddesses of the seasons

IO a water nymph, one of Zeus's lovers

ISHTAR the Mesopotamian goddess of love, warfare and justice

ISIS the Egyptian mother goddess of the afterlife, rebirth, magic and healing, a sky goddess

JASON a legendary Greek hero who sailed with the Argonauts in search of the Golden Fleece

KHEPRI the Egyptian god of the morning sun

MUSES the nine muses, goddesses of the arts: astronomy, comedy, dance and choral song, epic poetry, love poetry, sacred poetry, history, music, tragedy

NEPHELE a cloud nymph or goddess

NEREIDS Greek sea nymphs

NERGAL the Babylonian god of plague and pestilence

ORION a Greek huntsman and giant, said to have had three divine fathers: Zeus, Poseidon and Hermes

ORPHEUS a legendary Greek hero, famed as a musician, son of the Muse Calliope and King Oeagrus

PABILSAG a Babylonian god of war and hunting

PAN the Olympian pastoral god of shepherds and flocks, wild animals, bees, wild places, hunters and hunting

PERSEUS a legendary Greek hero, son of Zeus and the mortal Danaë, married to Andromeda, princess of Ethiopia

PHILOMELUS an Olympian demi-god who invented the plough; also identified with the herdsmen Boötes

POLLUX twin of Castor, joint gods of St Elmo's Fire (a faint light caused by an electrical discharge occurring on pointed objects such as ship's mast) and horsemanship; patron gods of sailors, guests and travellers, and the Olympic Games

POSEIDON the Olympian god of the sea, earthquakes and horses, husband to Amphitrite

PROMETHEUS the Titan god of fire, said to have created humankind from clay

RA the Egyptian god of the sun

RHEA the queen of the Titans, wife of Cronus

SEMELE the mortal mother of the god Dionysus, granted a home among the gods as a mother of a deity

SHAMASH the Babylonian god of the sun

SODPET the Egyptian goddess of fertility and prosperity

TITANS the race of giant gods who ruled the cosmos before the Olympian gods

VENUS the Roman goddess of love, beauty and fertility, also goddess of gardens, wife to Hephaestus and mother to Cupid

ZEUS the king of the Olympian gods, god of the sky and weather, and of law, order and justice, husband to Hera

Bibliography

SELECT BIBLIOGRAPHY

ADAIR, JAMES., *Rowing after the White Whale: A Crossing of the Indian Ocean by Hand* (Edinburgh, Polygon, an imprint of Birlinn Ltd, 2015)

ANDREWS, TAMRA., *Dictionary of Nature Myths: Legends of the Earth, Sea, and Sky* (Oxford, Oxford University Press, 2000)

BARENTINE, JOHN C., *Uncharted Constellations: Asterisms, Single-Source and Rebrands* (Springer Praxis Books, 2016)

BURRITT, ELIJAH HINSDALE., *The Geography of the Heavens, and Class Book of Astronomy, enlarged and revised by Hiram Mattison* (New York, Sheldon and Company 1860)

CAPINERA, JOHN L., *Encyclopedia of Entomology, 2nd edition* (Springer Science & Business Media, 2008)

COOK, ALAN H., *Edmond Halley: Charting the Heavens and Seas* (Oxford, Clarendon Press, 1998)

HESIOD, 'THEOGONY'., *The Homeric Hymns and Homerica with an English translation by Hugh G. Evelyn-White* (Loeb Classical Library: Cambridge, MA, Harvard University Press; London, William Heinemann Ltd, 1914)

HYGINUS., *Fabulae and Book 2 of Poetica Astronomica: The Myths of Hyginus, translated and edited by Mary Grant* (Lawrence, Kansas, University of Kansas Press, 1960)

OGILVIE, BRIAN W., *The Science of Describing: Natural History in Renaissance Europe* (Chicago and London, University of Chicago Press, 2006)

NORRIS, RAY P., *Dawes Review 5: Australian Aboriginal Astronomy and Navigation* (Publications of the Astronomical Society of Australia, Volume 33)

OVID., *Metamorphoses, with an English translation* by A. D. Melville (Oxford, Oxford University Press, 1986)

PENPRASE, BRYAN E., *The Power of Stars: How Celestial Observations Have Shaped Civilization* (Springer Science & Business Media, 2010)

PETERSON, DALE., *Giraffe Reflections* (University of California Press, 2013)

PROTHERO, DONALD., R A SCHOCH, M. ROBERT., *Horns, Tusks, and Flippers: The Evolution of Hoofed Mammals* (Baltimore and London, The Johns Hopkins University Press, 2002)

SIMPSON, PHIL., *Guidebook to the Constellations: Telescopic Sights, Tales, and Myths* (Springer Science & Business Media, 2012)

THEOCRITUS., *'The Poems of Moschus',* *The Greek Bucolic Poets, with an English translation* by J. M. Edmonds (Loeb Classical Library: London, William Heinemann; New York, G.P Putnam's Sons, 1912)

SELECT WEBSITES & ONLINE PUBLICATIONS

Ancient History Encyclopedia: **ancient.eu**

Ancient Mesopotamian Gods and Goddesses (AMGG) – based on data prepared by the HEA-funded AMGG project: **oracc.museum.upenn.edu/amgg/index.html**

Astronomy Trek – astronomy news and reviews, including a comprehensive guide to star constellations and stars: **astronomytrek.com**

Atsma, Aaron J., The Theoi Project – comprehensive profiles of gods, heroes, heroines and creatures of Greek mythology, including quotations from ancient Greek and Roman texts and a classical texts library: **theoi.com**

Colavito, Jason., Jason and the Argonauts Through the Ages – comprehensive coverage of the story of Jason and the Argonauts, including original texts: **argonauts-book.com**

Constellation Guide – comprehensive guide to the night sky: **constellation-guide.com**

Diodorus Siculus., The Library of History, with an English translation by C. H. Oldfather: **penelope.uchicago.edu/Thayer/E/Roman/Texts/Diodorus_Siculus/home.html**

Dolan, Chris., 'The Constellations and Their Stars', Department of Astronomy, University of Wisconsin-Madison: **astro.wisc.edu/~dolan/constellations/constellations.html**

Encyclopaedia Britannica: **britannica.com**

Encyclopedia of Art: **visual-arts-cork.com**

Glenn, Jacob., 'Divine Sky: The Artistry of Astronomical Maps', Shapiro Science Library, The University of Michigan: **lib.umich.edu/divine-sky-artistry-astronomical-maps/atlases.html**

Hall, Linda., 'Scientist of the Day – Jodocus Hondius': lindahall.org/jodocus-hondius/

International Astronomical Union 'Naming Stars': www.iau.org/public/themes/naming_stars/#table%7CList 'Constellation Names': iau.org/public/themes/constellations/

Kraft, Alexander., 'On the Discovery and History of Prussian Blue', Bulletin for the History of Chemistry, Volume 33, Number 2, 2008: scs.illinois.edu/~mainzv/HIST/bulletin_open_access/bull-index.php

Leeuwenhoek, Anton, van., 'Little Animals Observed in Rain – Well – Sea and Snow Water': commons.wikimedia.org/wiki/File:Leeuwenhoek-Little_Animals_Observed_in_Rain-Well-Sea.pdf

National Geographic: nationalgeographic.com

Nichols, Andrew., The Complete Fragments of Ctesias of Cnidus, 'The Indika' (Dissertation presented to the graduate school of the University of Florida, 2008): etd.fcla.edu/UF/UFE0022521/nichols_a.pdf

Parada, Carlos., The Greek Mythology Link: comprehensive resource on the Greek myths: maicar.com

Perseus Digital Library, Greek and Roman Materials: perseus.tufts.edu/hopper/collections

Ridpath, Ian., Star Tales: Myths, Legends and History of the Constellations: ianridpath.com/startales/contents.htm

Sea and Sky – includes a guide to star constellations, celestial objects, space exploration and astronomy news: seasky.org

Stuckey, Johanna., 'Atargatis, the Syrian Goddess': matrifocus.com/BEL09/spotlight.htm

TechnologyUK, 'A Brief History of Navigation Instruments': technologyuk.net/physics/measurement-and-units/navigational-instruments.shtml#ID06

Took, Thalia, 'Atargatis': thaliatook.com/OGOD/atargatis.php

Top Astronomer – astronomy and night sky guide for the UK: topastronomer.com

University of the West of England, Bristol, 'Georgian Architecture – Theory of Design': fet.uwe.ac.uk/conweb/house_ages/flypast/print.htm

Warner, Brian., 'Lacaille 250 Years on', Astronomy & Geophysics, Volume 43, Issue 2, 1 April 2002, Pages 2.25–2.26: academic.oup.com/astrogeo/article/43/2/2.25/281196

Wikipedia: wikipedia.com

Wright, Anne., Constellation of Words – explores the etymology and symbolism of the constellations: constellationsofwords.com

RESOURCES
ASTRONOMY WEBSITES

UK: An interactive sky chart to the stars over the UK: astronomynow.com/uk-sky-chart/

Dark Sky Discovery, online resource to the night sky: darkskydiscovery.org.uk/the_night_sky/

British Astronomical Association: britastro.org/community

Astronomy Now, astronomy magazine: **astronomynow.com**

Sky at Night Magazine, online astronomy: magazine: **www.skyatnightmagazine.com**

USA: Amateur astronomy clubs of the USA: **go-astronomy.com/astro-club-search.htm** Astronomy Magazine, an astronomy magazine: **astronomy.com**

Sky and Telescope, an astronomy magazine: **skyandtelescope.com**

Hawaiian Astronomical Society, home of amateur astronomy in Hawaii: **hawastsoc.org**

Canada: Royal Astronomical Society of Canada: **rasc.ca**

SkyNews, the Canadian Magazine of Astronomy & Stargazing: **skynews.ca/ resources/astronomy-clubs/**

Australia: Astronomical Society of Australia: **astronomy.org.au/amateur/amateur-societies/australia/**

Monthly sky guides: **maas.museum/ observations/category/monthly-sky-guides/**

FURTHER READING

DICKINSON, TERENCE., *The Backyard Astronomer's Guide* (Richmond Hill, Firefly Books, 3rd revised edition, 2008)
A guide for the home astronomer, including advice on telescopes and other equipment, astrophotography, star charts and atlas of the Milky Way

DUNLOP, STORM., AND TIRION, WIL., *Guide to the Night Sky* (London, HarperCollins Publishers) – a yearly publication for the skies above Britain and Ireland

ELLYARD, DAVID AND TIRION, WIL., *The Southern Sky Guide* (New York, Cambridge University Press 3rd Edition, 2008)

GATER, WILL, WITH GILES SPARROW., *The Night Sky Month by Month* (London, Dorling Kindersley Limited, 2011) Covers constellations in both the northern and southern hemispheres

OWENS, STEVE., *Stargazing for Dummies* (Chichester, John Wiley & Sons Ltd, 2013) Covers constellations in both the northern and southern hemispheres

POPPELE, JONATHAN., *Night Sky: A Field Guide to the Constellations* (Cambridge, MN, Adventure Publications, 2010) Covers most of the continental United States and southern Canada

PLANISPHERES

A planisphere, or star wheel, is a handy portable tool for locating constellations. It consists of two disks. The base disk depicts a star chart for your region, with the months marked around the edge. The second disk is a rotating overlay which is partially masked and has hours marked around the edge. Match the month with the time you are viewing the night sky and the clear window in the overlay disk will reveal the constellations that are visible.

UK AND CANADA ROYAL OBSERVATORY GREENWICH, DUNLOP, STORM, AND TIRION, WILL, COLLINS., *Planisphere: suitable for the UK, Ireland, Northern Europe and Canada,* latitude 50° North (Collins Maps, 2013) USA and other countries David Chandler Night Sky Planisphere – charts are available for a range of latitudes: **davidchandler.com**

ASTRONOMY APPS

STAR WALK 2, Vito Technology Inc
(Android, free; iOS) One of the most
popular stargazing apps, and includes lots
of features. Hold the device to the night
sky and it will identify constellations, stars
and planets.

NIGHT SKY, iCandi Apps (iOS, free)
Suitable for iPhone, iPad or Apple Watch:
hold the device to the night sky and it will
identify constellations, stars and planets.

NIGHT SKY LITE, iCandi Apps (Android,
free) Suitable for Android phones and
tablets: hold the device to the night sky
and it will identify constellations, stars
and planets.

SKY MAP, Sky Map Devs (Android, free)
Handheld planetarium for Android devices
GoSkyWatch Planetarium, GoSoftWorks
(iOS) Handheld planetarium suitable for
iPhone, iPad and iPod touch.